Discovery and Revelation

Discovery and Revelation

Religion, Science, and Making Sense of Things

Peter Manseau and
Andrew Ali Aghapour

Smithsonian Books
Washington, DC

Published by Smithsonian Books
Director: Carolyn Gleason
Senior Editor: Jaime Schwender
Assistant Editor: Julie Huggins

Edited by Tom Fredrickson
Designed by Studio A
Image research by Lauren Safranek

This book may be purchased for educational, business, or sales promotional use. For information, please write:

Special Markets Department
Smithsonian Books
P.O. Box 37012, MRC 513
Washington, DC 20013

Library of Congress Cataloging-in-Publication Data

Names: Manseau, Peter, author. | Aghapour, Andrew Ali, author.

Title: Discovery and revelation : religion, science, and making sense of things / Peter Manseau, and Andrew Ali Aghapour.

Description: Washington, DC : Smithsonian Books, 2022. | Includes bibliographical references and index. | Summary: "An illustrated history of how scientific study and religious thought have influenced each other throughout American history"—Provided by publisher.

Identifiers: LCCN 2021003621 | ISBN 9781588347015 (hardcover)

Subjects: LCSH: Religion and science—United States—History.

Classification: LCC BL245 .M36 2022 | DDC 215.0973—dc23

LC record available at https://lccn.loc.gov/2021003621

Printed in Singapore, not at government expense
26 25 24 23 22   1 2 3 4 5

Frontispiece: Bill Anders, Lunar Module pilot on Apollo 8, took this photo, aptly named *Earthrise*, on December 24, 1968.

Contents

Introduction

When Benjamin Franklin famously tied a key to a kite one stormy afternoon in 1752, he had no idea he was courting theological controversy. Franklin had a scientific goal: to demonstrate the "sameness of the electric fluid with the matter of lightning." With the help of his son William, he wanted to prove that lightning was electric.

Benjamin Franklin Drawing Electricity from the Sky by Benjamin West, ca. 1816.

They walked toward the storm. Franklin had affixed a small length of wire to the top of the kite to act as a lightning rod. An electric current, Franklin reasoned, would carry over the wire and the kite's wet hemp string, and the current could then be captured in a device known as a Leyden jar.

Despite some Founding Fathers myths suggesting otherwise, Franklin's kite was not struck directly by lightning. It did, however, successfully manage to draw an ambient electric charge from the skies, effectively proving the relationship between electricity and the flashes that emerged from thunderclouds. Franklin wrote in his instructions to any souls brave or foolish enough to try the experiment themselves:

When the Rain has wet the Kite and Twine so that it can conduct the Electric Fire freely, you will find it stream out plentifully from the Key on the Approach of your Knuckle. . . . Electric Fire thus obtain'd, Spirits may be kindled, and all the other Electric Experiments be perform'd, which are usually done by the Help of a rubbed Glass Globe or Tube; and thereby the Sameness of the Electric Matter with that of Lightning compleatly demonstrated.

He had literally captured lightning in a bottle. Brushing his bare skin against the charged key, it was as if Franklin had bridged the gulf between heaven and Earth no less than the eternally reaching fingers on the Sistine Chapel's ceiling.

George Washington Carver called his laboratory "God's little workshop" and saw it as a place where he carried out divine work.

During the time of Franklin's experiments, many American colonists believed lightning to be a supernatural force caused by either God or demons. Lightning was particularly concerning because the tallest edifice in most New England towns was the church steeple. These structures were ideal lightning targets, and church bell ringers were often electrocuted. One year after his kite experiment, Franklin published instructions for installing grounded metal rods on rooftops in order to safely divert this "Electric Fire."

Some of Franklin's contemporaries sensed the religious implications of his efforts and did not appreciate them. To manipulate the elements as Franklin had done was to approach the blasphemous notion that human striving could supplant providential will. Who was Benjamin Franklin to presume humankind should harness and control a force that had been considered a symbol of divine wrath since Zeus wielded it on Mount Olympus? As one critic charged, "It is as impious to ward off Heaven's lightnings as for a child to ward off the chastening rod of its father."

Yet in constructing this early iteration of a lightning rod, Franklin had no such irreverent intentions. Precisely the opposite: As a student of what was then called natural philosophy, Franklin believed that increased knowledge of the world and its workings could only provide greater understanding of God. To grapple with the power of creation as he had, to apply his limited human mind to the omnipotent mind that he believed moved the universe—how could such pursuits inspire anything but purest awe? As for what aroused suspicion of lightning rods, Franklin's friend John Adams had some piquant thoughts: "This Invention of Iron Points to prevent the Danger of Thunder," he wrote in 1758, "has met with all that opposition from the superstition, affectation of Piety, and Jealousy of new Inventions, that . . . all other usefull Discoveries have met with in all ages of the World."

The surprising theological scandal that flared around Franklin's experiments with electricity was not the first contretemps between religious ideas and scientific thinking in American history, and, of course, it would not be the last. Decades before, colonial Massachusetts had been roiled by disagreements over the proper treatment of smallpox, with some Puritan ministers arguing for the new practice of inoculation and some against, each faction confident the will of God was on their side. Two centuries later, the infamous Scopes trial electrified the nation with an almost theatrical legal debate over the teaching of evolution in public schools.

And yet from the very beginning of the American experiment—at least since the long period of exchange of European and Indigenous knowledge, as when Wampanoag peoples shared ecological expertise with their Puritan neighbors and began to make use of colonial technologies

Etching of George Washington Carver by Felix B. Gaines, 1946. Carver was a prominent scientist and leading environmentalist who believed that the mysteries of God could be studied in nature.

in return—science and religion have also been entwined in ways both practical and profound. The nation's diverse spiritual traditions have often been at the forefront of new technologies or the first to embrace important scientific ideas and incorporate them into a shared cosmology.

American scientists have also come from varied backgrounds and worldviews, bringing to scientific practice a surprising range of religious and spiritual notions. Botanist George Washington Carver, who developed techniques to improve soils that had been depleted by generations of cotton production, called his laboratory "God's little workshop" and saw it as a place where he carried out divine work. The Nobel Prize–winning geneticist Barbara McClintock considered herself to be a kind of scientific mystic who was able to make her discoveries of genetic transposition through a transcendent form of understanding. "The point is," she said of science's tendency to divide into subfields and separate researchers from their subjects, "we forget we are all one thing." When the Columbia University physicist Charles Townes conceived of a device to produce coherent electromagnetic waves through amplification of stimulated emission of radiation—the basis of the laser—he considered it less a discovery than a revelation, likening it to a gift from God. Albert Einstein,

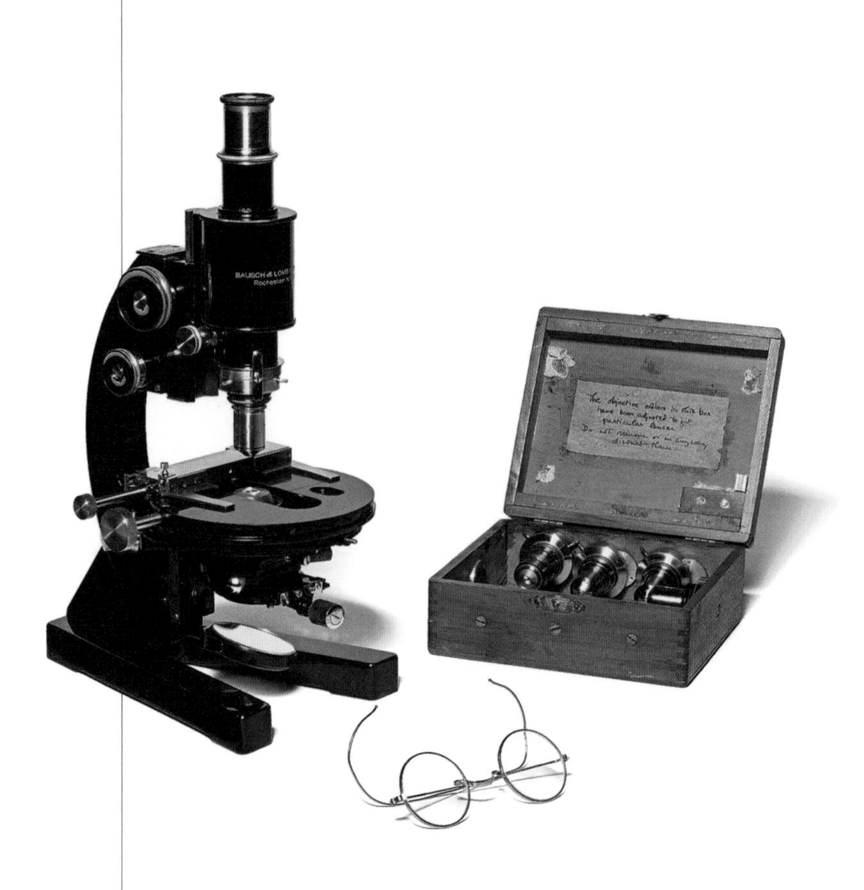

Microscope and other laboratory materials used by Barbara McClintock, ca. 1940. McClintock used the instrument to study the phenomenon of "jumping genes," or segments of DNA that move between locations on a chromosome. She likened science to a kind of mysticism.

on the other hand, rejected the notion of a personal God, arguing that no omnipotent being would allow for the existence of evil and suffering. Still other American scientists have sought to place religion itself under the microscope—or brain scanner—and observe it scientifically.

When we attend to the lived experiences of Americans over the past 400 years, we are also reminded that religion and science are not merely abstract categories; they are also the stuff of everyday life. Religion and science deeply influence people's identities, opportunities, imaginations, communities, and worldviews. We see this, for example, in Harriet Powers's Bible quilt (see page 90), which stitches together Bible stories, the spirituals of enslaved people, and astronomical observations into a single fabric. Whether in harmony or discord, religion and science together spark the discoveries and revelations that shape all of our lives.

This book, like the *Discovery and Revelation* exhibition at the National Museum of American History upon which it is based, considers the dynamic interplay of religion and science in American history. For centuries, their tensions and syntheses have made for a nation that is both religiously and scientifically vibrant. Rather than telling a story of either inevitable conflict or uncomplicated agreement, the objects, images, and stories presented here examine the mutual influence of two ways of knowing that continue to shape the American experience.

According to a 2015 study by the Pew Research Center, a majority of Americans believe that science and religion are frequently in conflict. Survey findings over time suggest the percentage of those holding this opinion is on the rise, with 59 percent agreeing to this supposed clash, up from 55 percent just five years before. Yet interestingly, when it comes to one's own beliefs, most see no sign of incongruity. When asked "Does science sometimes conflict with your own religious beliefs?" more than two-thirds of respondents said unambiguously that it did not. Apparently, some Americans take for granted that religion and science are at odds despite their ability to reconcile them in their own worldviews.

Some Americans take for granted that religion and science are at odds despite their ability to reconcile them in their own worldviews.

One year after the Pew study, curators and educators at the Smithsonian became curious how answers given to a national survey would track with the experiences of visitors to the National Museum of American History in Washington, DC. We spoke informally to museum goers from across the country and around the world to find out what the intersection of religion, science, and the latter's practical application through technology brought to mind. The responses we heard—anecdotal, idiosyncratic, thoughtful, sometimes humorous—helped personalize the data that had inspired our conversations.

Many subjects that came first to mind for visitors tracked neatly with the assumption of conflict that broader surveys have documented. When asked what words they associated with religion and science, most visitors responded with words like "conflict," "contradict," "opposites," and "controversy." One visitor hoped to learn more about "the Scopes trial and teaching evolution in school," while another asked, "How was the universe created? The Bible says one way, science says another." More than one person told us "There is always a fight between science and religion."

Many began with black and white extremes but then, as they considered the topic further, discovered areas of gray. "What is the scientific side of religion?" one visitor asked. "Where is it in the brain?" Others immediately seemed comparatively kaleidoscopic. One guest said that religion and science spoke to "our relationship to the natural world" and hoped to learn more about "religions that worship plants and trees." Another asked, somewhat dreamily, "Have scientists tried to prove the existence of God?"

A few of our conversations with visitors reminded us that the topic of religion and science is not limited to abstract intellectual debates. A rabbi told us that the topic of religion and science "always comes up" in her community, since both are viewed so often as sources of good and evil. A teenager said that he wanted to bring his mother to the exhibition so that they could have a "big conversation" concerning matters of faith and fact where their opinions diverged. One visitor, a Catholic school teacher, offered candid advice: "Religion and science can mix, but that's a touchy subject."

As the range of responses from visitors suggests, both science and religion are expansive topics, and they are often slippery as a result. For our purposes, we have sought not to limit either but rather to embrace the ambiguity of each term's changing use across history. Insofar as we will define what is religious and what is scientific, we will err on the side of allowing each to be more than many might assume.

If the intersection of these two hard-to-define categories is indeed a "touchy subject," it may be because each on its own seeks to address the big questions of life. Particular expressions of these questions are as numerous as the people who ask them. *Discovery and Revelation* was organized around American efforts to offer answers to just three questions broad enough to address a great variety of concerns: *Who are we? Where are we?* And *How should we live?* These three thematic lenses allow an exploration of how answers to existential quandaries have changed over time, while noting also that the drive to ask such questions has been an engine of change throughout our history.

Who Are We?

Science and religion are both well equipped to offer possible answers to humanity's ultimate existential questions: What does it mean to be human? What is unique about our species and our consciousness? What are our bodies made up of? Where did we come from?

Science and religion are both well equipped to offer possible answers to humanity's ultimate existential question: What does it mean to be human?

The question of human origins has long been a flash point between religion and science. Great public debates since the nineteenth century have concerned the crucial questions of how species change and how we should understand the place of humanity within the broader spectrum of life. At the same time, religious traditions have evolved in response to evolutionary theory, in some cases becoming more recalcitrant in the face of the supposed assault of science (leading to the rise of fundamentalism and biblical literalism), but in others incorporating new scientific understandings into religious self-identity.

In *Discovery and Revelation* this story is told through objects related to the Darwinian debates beginning in the 1860s as well as the popular culture of the 1920s onward, when the image of Darwin became a significant symbol driving the perception of a divide between scientific and religious worldviews. The evolution debate also played a significant but often forgotten role in Americans' understanding of social issues, including race. Darwinism had challenged alternative, biblically inspired theories of human origins that insisted different racial groups had been created separately by God, providing justifications for slavery and, later, racist social policies. Over the course of American history, scientific and religious origin stories alike have been used to bolster—and challenge—social hierarchies, with unfortunate implications, as we shall see.

Who are we? is a question about human bodies, but it pertains to more than just flesh. To examine the history of asking who we are, one must also consider America's perennial obsession with the science of the soul. From spirit detectors and phrenological models of the nineteenth century to brain-scan imaging of the twenty-first, Americans have often turned to technologies that promise objective measurements of an otherwise ephemeral human spirit.

Where Are We?

Where are we? is a question not just about our location or coordinates, but about the fundamental nature of the universe that surrounds us. For early scientists, there seemed to be no discord between religion and science on the topic of where we are: the "book of Nature," they believed, could be studied alongside the "book of Scripture" to reveal the creative works of God.

Over time, however, the worlds depicted by these two books would diverge. Throughout the eighteenth century, the improved ability to peer into "the heavens" afforded by ever more powerful telescopes also offered Americans a better view of the expansive universe of which the Earth was but a tiny part. During the nineteenth century, geological discoveries concerning the age and composition of the Earth radically remade popular assumptions about the history and origins of our planet and the vast universe beyond.

For some, these new scientific accounts of the universe seemed incommensurable with scripture. Thomas Paine, the English-born American philosopher and revolutionary, provides a particularly vivid example of how science can transform one's personal religious views. Paine suggested that the orrery, a three-dimensional representation of the solar system, prompted his conversion from traditional Christianity to Deism, a belief in an eternal but indifferent God who plays no role in human affairs.

While Paine supposed that no one could see the true scale of the universe and preserve traditional beliefs, this was not the case for many Americans who later studied the stars. In the nineteenth century, astronomy became more widely practiced in part because many believed it could reveal information about the nature of divine creation. In the twentieth century, it was not considered out of place for the astronauts of Apollo 8 to broadcast words from the Bible as they orbited the moon.

How Should We Live?

While the massive intellectual disruptions surrounding evolutionary theory and the age of the Earth spurred near-universal changes to American religious thinking, more localized intersections of religion, science, and technology have had practical influence on many other aspects of American life, including birth and death, healing and education. Across a diverse range of contexts, bodies, and experiences, Americans have used religion and science to ask one of the fundamental questions of democracy: *How should we live?*

Perhaps most significantly, issues surrounding the beginning and end of life were flash points between religion and science in the nineteenth and twentieth centuries. Biblical notions of the supposed need for women to suffer during childbirth were once routinely discussed alongside techniques to ease the pain of delivery. Contraception and artificial insemination have provided fodder for debates surrounding how and when medical science should intervene in a process that some religious viewpoints regard as sacred. At the other end of life, American religious practices surrounding death have also changed with science and technology. The growth of modern chemical embalming during the Civil War especially changed attitudes toward the bodies of the dead; at around the same time, new religious movements like Spiritualism, the belief that the souls of the dead could be contacted from the realm of the living, adapted scientific practices such as the use of electric conduction in efforts to prove the immortality of the soul.

How should we live? is also a question raised at the intersection of public health and personal belief. Long before contemporary vaccination disputes, for example, Puritans debated whether smallpox inoculations played a role that should be reserved for God. From the creation of immortal cell lines to the scientific support of the Native American sacrament of peyote as medicine, science has played an intimate and sometimes sacred role in American life.

Together, these three question-driven themes provide an entryway for considering a history that might otherwise seem overly abstract or obscure. Yet the most important question a history museum can ask is "How have the answers to these questions changed over time?"

We have framed this book chronologically, organizing some of the objects, images, and stories of the *Discovery and Revelation* exhibition into three sections each covering roughly a century. The first, "Revolutions," explores the tumultuous eighteenth century, a period marked by the American War for Independence and the growth of Enlightenment ideas on American soil. The second section, "Evolutions," covers a period that was likewise a time of dramatic change. During the nineteenth century, Americans reckoned with evolutionary theory and an increasingly material view of the cosmos. It was during this era that the notion of an inherent conflict between religion and science first arose, and many Americans became firmly split over which deserved cultural and intellectual authority. Despite this apparent rift, however, religion and science would often prove to be enduringly entwined. The third section, "Complexity," explores the surprising places where religion and science could be found interacting during the twentieth century, from cancer cells to the surface of the moon. All of the narratives that follow explore instances—some infamous and some little known—when the complexity of religion, science, and technology came to the fore in American history.

Finally, a methodological coda: The intersections of religion and science are almost infinite, and this is no random sample. In this book we have attended to people and objects that exemplify certain motifs and paradoxes in the American story. In the gathering of objects and the telling of stories for this project, care has been taken to represent both the diversity of American religion and the heterogeneity of the American sciences. We have sought to calibrate against well-known biases, including the underrepresentation of women in the history of science, for example, and the overrepresentation of belief as the primary element of religion. We have also been keenly aware that the telling of stories based on enduring historical objects can bring its own distortions, privileging those with the resources and predilection to preserve their material history.

Given the longstanding demographic dominance of various forms of Christianity in the United States, a large portion of the examples presented here involve efforts to reconcile Christian teachings with new developments in geology, astronomy, biology, and other disciplines. The relationship between religion and science is not limited to one faith, however. The full range of spiritual perspectives to which the United States is home brings added dimensions to the relationship and is represented here as well.

Though grounded in history, the stories and objects gathered here also look forward to the next interactions of religion and science as each continues to change. By examining the subject both chronologically and thematically, *Discovery and Revelation* seeks to highlight any patterns that may appear, which in turn could provide new ways of approaching moments of conflict and congruence that we cannot yet imagine. When Benjamin Franklin's yet-to-be-born successors capture the future's version of lightning in a bottle, they will be better equipped to grapple with its scientific and spiritual significance if the history of such discoveries has been well revealed.

Issues surrounding the beginning and end of life were flash points between religion and science in the nineteenth and twentieth centuries.

REV
OLU
TIONS

The Scientific Revolution and the American Revolution each helped create a nation where the relationship of religion and science would be informed equally by disagreement and influence. Some spiritual traditions fostered education and inquiry, while scientific approaches to fundamental questions transformed religious views. Yet tension between belief in a God who intercedes in history and Enlightenment notions of a "clockwork universe" set the stage for future conflict between scientific discoveries and religious truth.

The encounter of European intellectual trends with knowledge of Indigenous and enslaved populations helped make early America home to a disruption of traditional understandings of authority, allowing science the opportunity to offer answers that once belonged to religion alone. During this period, religion and science were not widely believed to be distinct. Many revolutionary thinkers combined religious concepts and scientific experiments as they created and imagined new worlds.

Well-known figures, including Cotton Mather and Thomas Paine, along with lesser-known, like clockmaking astronomer Benjamin Banneker and the chemist nun Francis Xavier Hubert, shaped discourse surrounding religion and science in ways that can still be observed. Discoveries about the movement of the planets, the composition of the air we breathe, and the treatment of illness each presented new venues for reconsidering sources of knowledge. From assumptions of the complementary nature of spiritual seeking and experimental methods to the first stirrings of mutual mistrust, the colonial period and the early republic established enduring terms of engagement for religious and scientific ideas.

Automaton of a friar (detail), sixteenth century. This mechanical monk performed the gestures of prayer, raising questions about the practice and content of ritual.

The Mechanical Monk

Automaton of a friar, sixteenth century.

If a single object were to capture the creative tension that existed between religion, science, and technology in the era from which America emerged, this mechanical monk might raise its hand to be noticed. While on first inspection it appears the very image of conventional religious doctrine, practice, and authority—a Catholic friar praying with rosary beads—a look just under the surface reveals turning gears evocative of a subtly disruptive worldview. The figure runs not by magic or faith but by clockwork.

Little is firmly known about the automaton the Smithsonian acquired in 1977. We know it stands fifteen inches in height and resembles a Franciscan friar. We know it is constructed of iron and wood. We know, based on the particularities of its craftsmanship—the use of pins and not screws, of iron and not brass—that it was made in Europe during the mid-sixteenth century. We know that when its key-wound spring is tightened and released, it walks in a trapezoidal pattern, striking its chest with one arm and lifting a wooden cross and rosary in the other. While pacing, the monk nods its head, rolls its eyes, and works its jaw, as if silently mouthing a prayer. It occasionally raises the cross to its cracked lips, which still contain traces of flesh-colored enamel.

The mechanical monk was a technological marvel when it was first built, one of the earliest known examples of a "self-acting" automaton whose gears and springs were completely hidden inside its body. Finally, we know this: that, built in a world roiled by the Scientific Revolution and the Protestant Reformation, this robotic monk was a walking, praying theological paradox. It—or perhaps he—was an object that prayed. Never before had there been a machine so lifelike as to question whether the essence of ritual was spirit or if the so-called soul might be replaced with cogs and gears.

Beyond these facts there are theories and legends. According to one, the automaton was originally commissioned in the 1560s by King Philip II of Spain in thanks to God for the life of his son. As the story goes, Philip's son Don Carlos fell down a flight of stairs and lost consciousness. After Philip prayed at his son's deathbed, Don Carlos miraculously recovered and reported encountering in a dream a Franciscan monk who fit the description of the deceased Spanish saint Diego de Alcalá. In deference to God, the story goes, the king commissioned renowned clockmaker Juanelo Turriano to build a mechanical likeness of the saint.

While this story cannot be verified, we can deduce a few additional facts about the mechanical monk based on how it was crafted. In her 2002 essay on the object, sculptor and author Elizabeth King interviewed clockmaker W. David Todd, who studied the automaton in his role as the Conservator of Timekeeping at the Smithsonian Museum of History and Technology (now the National Museum of American History). Todd scrupulously examined every detail of the monk, even using X-rays to study its inner secrets. Todd's analysis, King writes, reveals "the work of a master mechanic with a restrained sense of style." King continued:

> **The complex design of the monk's left arm, with the elbow moving independently of the shoulder, alone is worth respect, and here it is done with an elegance only God was meant to see. . . . Todd points out a hidden lever to be used secretly by the operator of the automaton: once wound, the machine would only begin moving after the release of this lever. . . . We must envision a scenario, Todd advises us, where a powerful person, or an emissary from that person, is seen to hold the miniature man in his hands, then set it down on the table or floor. Whereupon, very slowly, very deliberately, very irrevocably, it would set out on its own.**

The mechanical monk was designed, it seems, to astound. To further appreciate how it might have affected its viewers, we might consider the profound intellectual and social changes transforming Europe at the time of its construction.

In 1517 the German theologian Martin Luther wrote his Ninety-five Theses criticizing the Roman Catholic Church's practice of selling indulgences, or remissions of sin. Although its accuracy is debated among historians, the popular image of Luther nailing his theses to the church door evokes impact of Luther's ideas, which initiated a fracturing of Western Christianity that we now call the Protestant Reformation. Less than thirty years later, the astronomer Nicolaus Copernicus published his *On the Revolutions of the Heavenly Spheres*, which upended traditional cosmologies with the argument that the Earth revolved around the sun. Many historians cite Copernicus's 1543 work as the beginning of the Scientific Revolution.

"Between the sixteenth and eighteenth centuries," writes religion scholar Randall Styers, "Europe experienced massive economic and political transformations. This era saw the fracturing of religious unity, the consolidation of the nation-state, and the emergence of new capitalist economic structures. . . . European powers launched an extended program of discovery and conquest of the non-European world that produced not only new riches but also a startling array of information from missionaries and explorers."

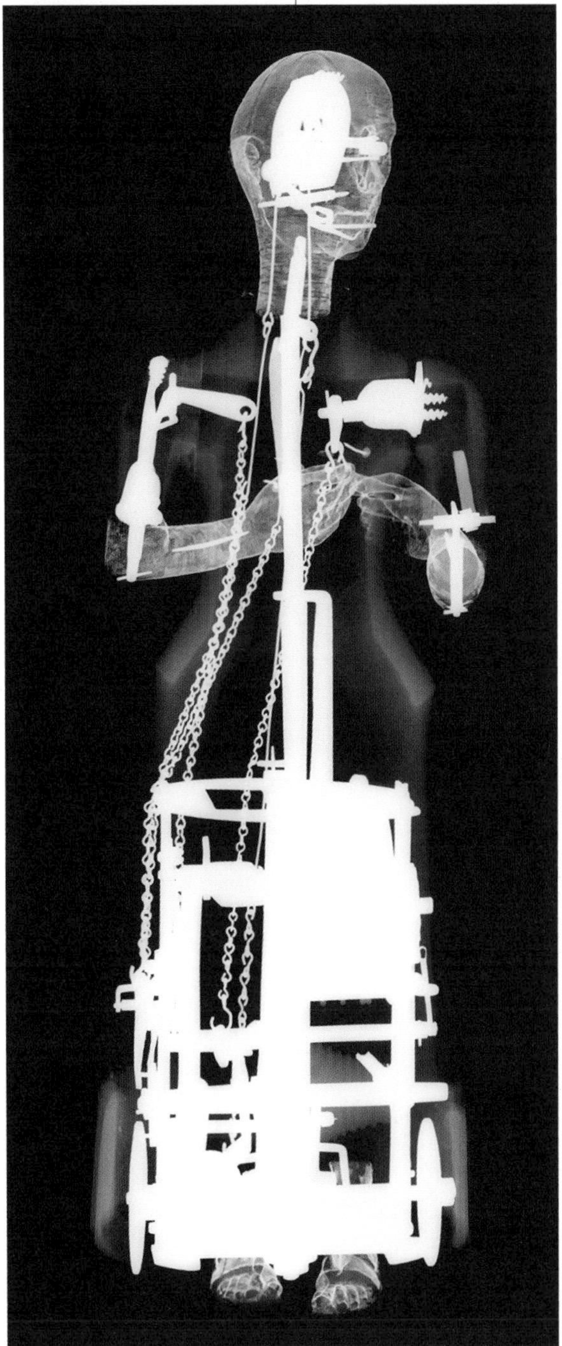

Thus, while the mechanical monk was being made, the modern world was being born. These social and intellectual changes would fundamentally transform European—and, later, American—understandings of religion. Where the word "religion" had once referred to the dutiful performance of ritual obligations, over the sixteenth and seventeenth centuries it increasingly came to be understood as a set of beliefs or an inner state of mind. Protestant reformers and Enlightenment thinkers alike viewed religious objects and rituals as "dubious encrustations" that ought to be relegated to history.

In this light, the mechanical monk was made of more than iron and wood. Among its inner workings lay coiled the central philosophical and theological questions of its day: How does religion occupy the material world? Is it wholly reducible to the stuff of matter, or is there something essential to religion that can't be replicated physically, no matter how precise the technology? Or, perhaps, can technology create religion anew? These questions, like the monk itself, would soon migrate to the United States.

X-ray showing internal mechanisms of automation of a friar.

Galilei

Ill.mo et Ecc.mo Sig.re e Pad.n mio Col.mo

La tra di V.S. Ill.ma et Ecc.ma sparsa tutta d'affetti di cortesia, e benignità, continua di farmi
parer sempre più soave la fortuna del mio infortunio, et in certo modo benedir le
persecuzioni de' miei nimici, senza le quali mi sarebbe sempre restata occulta la parte più
da stimarsi dell'humanità, e benigna propensione di molti miei Sig.ri e Pad.ri e sopra tutti
l'Amore di V.E: il quale nõ meritando d'esser promosso da talento alcuno di virtù che la
Natura habbia riposto in me, ha in vece di lei supplito la sorte cõ accender nelle lor menti
il fuoco della Carità, cõ la quale vanno compassionando lo stato mio; nel quale oltre alla
ragion detta mi è di nõ piccolo sollevam.to il creder che, non un animo che sempre più si
vada inasprendo sia quello che continui di tenermi oppresso, ma più presto una quasi dirò
ragion di stato di quelli che voglion ricoprir il primo errore d'haver a torto offeso un'innoce-
te, col continuar l'offese, e i torti, acciò l'universale si formi concetto che possano altri gravi demeriti
nõ fatti palesi aggravar la colpa del Reo. Hor sia quel che piace à chi è conceduta la potestà di
fare il suo arbitrio, che in tutti gl'eventi resterò io perpetuam.te obbligato alla somma bontà di V.E
la quale con tanta premura si appassiona nel mio interesse, e cõ tanta industria, e vigilanza
indefessam.te va specolando i mezzi che possano essermi di sollevam.to

L'Horologio Hydraulico sarà veram.te cosa di estrema maraviglia, quando sia vero che il Globo pendente nel mezzo dell'Acqua vada naturalm.te volgendosi p occulta virtù magnetica.
Io feci già molti anni sono una simile invenzione; ma cõ l'aiuto d'un ingannevole artifizio,
e la Machina era tale. Il Globetto diviso cõ 12. meridiani p le 24. hore era di rame, voto dentro,
e cõ un pezzetto di Calamita, postogli nel fondo, equilibrato quasi alla gravità dell'Acqua, si che
posta nel vaso una parte d'acqua salata, e poi sopra quella altra dolce, il Globo si fermava
tra le due acque, cioè nel mezzo del vaso: il qual vaso posava sopra un piede di legno dentro
al quale stava ascosto un Horologio fabbricato à posta con tal'arte che girava un pezzo di Calamita, che sopra vi era accomodata, facendogli fare una revoluzione i 24. hore al cui moto
ubidiva l'altra calamita posta nel Globetto facendolo girare, e mostrar le hore. Sin qui arri-
vò già la mia specolazione: ma se questa del P. Lino senz'altro artifizio fa che il suo Globo
ubbidisca al moto del Cielo, sarà veram.te cosa celeste, e divina, et haremo il Moto perpetuo.
V.E con quei mezzi che va nominando potrà facilm.te venire in cognizione del tutto, io tra tanto ne ho voluto significare il mio pensiero p havere un testimonio omni exceptione maius
che nõ ho usurpata l'invenzione al P. Lino; se però la sua machina nõ havesse altro di più
che la mia.

Non devo nasconder à V.E. come sentendo un Principe grande l'ordine mandato dal
S.to Off.o à tutti gl'Inquisitori di nõ dar licenza nõ solam.te che si ristampi alcuna
delle opere mie già molti anni fa pubblicate, ma che nõ si licenzij alcuna di nuovo

And Yet It Moves

Galileo Galilei (1564–1642) never crossed the Atlantic, but his figure loomed large in the thirteen colonies and the young United States. As American ideas about the relationship between science and religion developed throughout the eighteenth century, the memory of the Renaissance mathematician and inventor provided a model for both those seeking new discoveries and those who feared their inquiries would meet theological opposition.

Letter from Galileo Galilei to Nicolas Peiresc, 1635. Written from under house arrest during the Inquisition, the letter documents Galileo's fear that his scientific discoveries might be suppressed.

Thomas Jefferson and other Enlightenment figures used Galileo's 1633 trial by the Roman Inquisition for publishing works in support of heliocentrism (the understanding that the sun is the center of the solar system and the planets revolve around it) as an example of religious—specifically Catholic—hostility to Enlightenment thinking. Through the centuries that followed, Galileo remained a potent symbol, standing at once for science itself and for the refusal to back down in the face of religious authority. Over time, he became for many an embodiment of the view that religion and science are in inevitable conflict.

Galileo's story is more complicated than the uses to which it has been put, however. The writing that earned him the church's rebuke was initially undertaken with the blessing of none other than the pope. The punishment he received was evidence not only of religious intransigence but of the quickly shifting sands of Renaissance politics.

Born in the Italian city of Pisa in 1564, Galileo in his youth studied medicine and mathematics, becoming a professor of the latter in his twenties. Yet he never lost the budding medical student's penchant for working with his hands. While teaching at the University of Padua, Galileo designed and built such practical tools as a geometric and military compass and, most fortuitously, a refracting telescope, which allowed him to make the systematic celestial observations that would change his life.

che io, od altri volesse stampare, sì che la proibizione è de omnibus editis, et edendis, si è preso assun-
to di fare stampare il resto delle mie fatiche non publicate ancora, e forse si è mosso per curiosità
di veder l'esito di questa impresa, e che fortuna correranno tali materie contaminiss: da proposiz.
attenenti à religione più che non è il Cielo dalla Terra. Io contro à mia voglia sono stato forzato à con-
cedere copia à S. A. sicuro che à me non ne possa succeder se non qualche travaglio; se bene non mi è
stata fatta, ne accennata proibizione alcuna; perchè non devo ne anco haver notizia del divieto fat-
to à gl'Inquisitori; per lo che questo che conferisco à V.E. sia detto in confidenza. Da questo, e dall'esser
stati raccolti in Firenze, et in Roma tutte l'opere mie sì che più non se ne trovano per le librerie, apertam.
si scorge che si fa ogni opera per levar dal mondo la mia memoria; nella qual vanità, se sapessero i miei avversarij quanto poco io premo, forse non si mostrerebbero tanto ansiosi d'opprimermi.
Io non finirò di parlar con lei senza di nuovo ringraziarla della sua infinita benignità, e del fervore
col quale interceda ne miei interessi, e se il sollevare chi fuor di tutte sue colpe viene travagliato
è atto meritorio, può V.E. viver sicura che ne riceverà guiderdone dalla Divina bontà. E qui con reverente affetto gli
bacio le mani, e nella sua buona gra. mi rac.e
Dalla Villa d'Arcetri li 12 di Maggio 1635

Di V. S. Ill.ma et Ecc.ma

Dev.mo Et Oblig.mo Ser.re
Galileo Galilei

Letter from Galileo Galilei to Nicolas Peiresc (continued from page 22), 1635.

Through his lenses, Galileo was able to see such wonders as Saturn's rings and the moon's surface. At the time, the prevailing understanding of celestial objects was that they were higher order creations and thus were perfectly smooth. Nonetheless, when he peered at the moon, he found it to be, as he wrote, "uneven, rough, and crowded with depressions and bulges." Complicating received wisdom soon proved to be the primary role assigned to him by history.

Galileo did not single-handedly overturn tradition with his novel ways of thinking but rather affirmed theories made before he was born. In 1543 the Polish polymath Nicolaus Copernicus had challenged the widely held belief that the Earth was still and all the planets revolved around it. This geocentric understanding of the universe, which dated to the Greco-Egyptian astronomer, mathematician, and geographer Ptolemy, had dominated astronomical understanding for centuries. Attempting to prove the Ptolemaic theory mathematically, Copernicus found he could not. Rather than the Earth standing still, he suggested, it was more likely that it moved, along with all the other planets, around the sun.

As early as the 1590s Galileo expressed belief in the Copernican position, but due to an intellectual climate against novelty in this and other cosmological ideas, it would take him another two decades—and several significant discoveries with his telescope—to speak openly about it. Even then, the stakes were daunting. In 1600 his fellow stargazer Giordano Bruno was burned at the stake for loudly espousing the views Galileo more prudently kept to himself.

A decade after Bruno's death, Galileo built his most powerful telescope yet. With it, he was able to view the moons of Jupiter, whose orbit around the massive planet confirmed the Copernican model. Heavenly bodies did indeed move, and they did so on a nearly circular path, without the Earth as their hub. Thereafter he spoke more openly about heliocentrism, while simultaneously hearing from high-ranking officials in the church that he should be cautious about sharing such teachings.

Yet the highest ranking official of all, Pope Urban VIII, encouraged Galileo to publish a work considering both theories of celestial movement side by side. In 1632 Galileo published his *Dialogo sopra i due massimi sistemi del mondo* (Dialogue concerning the Two Chief World Systems), presented as a discussion between advocates of Copernican heliocentrism and Ptolemaic geocentrism. Though tasked with not taking sides in his presentation of this dialogue, Galileo could not resist giving heliocentrism the upper hand. Moreover, in a potentially fatal artistic miscalculation, he enlisted some of the pope's own words on the subject in the dialogue. Though Urban had asked for his thoughts to be included and might have been flattered by such an homage, he was far less so when it was clear that Galileo had stacked the argument against him, presenting the pope's view as lacking in coherence.

At the trial that soon followed, Galileo was threatened with torture if he did not recant. Knowing the fate that had met Giordano Bruno, he publicly admitted to errors in his work, while privately grumbling that his opinions on the matter would never change. According to legend, after recanting Galileo immediately looked down to the Earth and uttered, "E pur si muove." *And yet it moves.*

The letter shown in this essay was written in Galileo's hand in 1635, two years into the house arrest that would continue until his death. Despite bearing the scrutiny of the Inquisition and the pique of the pope, Galileo continued to have influential defenders. One was Nicolas Peiresc, a French priest and astronomer who regularly petitioned the Vatican seeking clemency for his friend. Peiresc wrote in 1634 to Cardinal Barberini, a longtime acquaintance of Galileo and, more importantly, a nephew of Pope Urban VIII. He argued that failing to grant Galileo the measure of mercy his life's work warranted could become a blemish on the pope's reputation. Many now would agree that it has.

Much of Galileo's two-page letter to Peiresc concerns a newly designed water clock, which featured a submerged orb that seemed to spin endlessly with no outside force exerted upon it. Peiresc had described the clock to the great inventor as if it was a world-changing innovation whose perpetual motion would at last prove Copernican theory—and thus Galileo's claims. Galileo, however, had seen such clocks before. He understood its inner workings and so knew not to put any faith in its ability to clear his name.

"The water clock will truly be a thing of extreme marvel if it is true that the globe suspended in the middle of the water goes naturally turning by an occult magnetic force," Galileo wrote. "Years ago I made a similar invention but with the aid of a deceptive artifice."

The Inquisition intended to fully silence Galileo by preventing the publication of troublesome ideas.

Having built one himself, Galileo gently informed his ardent defender that the clock's movement was likely explained by human ingenuity rather than measurable cosmic forces that would demonstrate he had been right all along. It would be no silver bullet bringing an end to his troubles.

The letter is of particular interest for its discussion of the limitations that had been placed on Galileo's work. As Galileo explained, his punishment restricted not only where he could go, but where and how widely his words might travel. The Inquisition intended to fully silence him by preventing the publication of troublesome ideas. At the time of this correspondence, the fate of all his writings remained an open question—not only those related to heliocentrism but even those, he said, "more remote from propositions pertaining to religion than are the heavens from the Earth." The ultimate goal, he feared, was to rid the world not merely of the notion that the Earth was not the center of the universe, but those who dared speak it. "It is easy to see that every effort is being made to remove all memory of me from the world, but if my adversaries knew how little I strive for such vanity, perhaps they would not show themselves so anxious to oppress me."

Portrait of Galileo Galilei by Galgano Cipriani, ca. 1800. Galileo was persecuted during the Roman Inquisition for his defense of heliocentrism. Centuries later, Enlightenment thinkers like Thomas Jefferson remembered him as a scientific martyr.

Galileo's letter to Peiresc is part of the collections of the Dibner Science Library at the National Museum of American History. While it did not join the story of religion and science in the United States until centuries after it was written, it nonetheless represents an element of their intersection with enduring influence in this country.

In spirit, the letter echoes Galileo's parting shot to the Inquisition—*And yet it moves*—which in the intervening centuries has become a slogan for the incontrovertibility of scientific fact even under duress. Some 250 years after the publication of his own heretical writings, the spirit of his words would serve as the central thrust of another controversial book, *History of the Conflict between Science and Religion*, published in New York by John William Draper in 1874 (see page 94). An ocean away and in an entirely different context, Galileo's memory was enlisted in the first salvoes of a cultural skirmish that no one then knew would help define the decades that followed.

INOCULATION
OF THE
Small POX
As practiſed in *Boſton*,
Conſider'd in a Letter to
A—*S*—M.D. & F.R.S.
In LONDON.

So learned Taliacotius *from*
The brawny Part of Porters Bum,
Cut ſupplemental Noſes——

Hudibras.

BOSTON:
Printed and ſold by *J. Franklin*, at his Printing-Houſe in Queen-Street, over againſt Mr. *Sheaf*'s School. 1722.

Something of Ye Small-Pox

While he was perhaps best known for his writings on demonic possession that helped set the stage for the Salem Witch Trials in 1692, the Massachusetts Puritan minister Cotton Mather (1663–1728) is also an important figure in the American history of science and medicine.

Mather's pathbreaking 1721 book *The Christian Philosopher: A Collection of the Best Discoveries in Nature, with Religious Improvements* included short essays on a range of topics then in the purview of the discipline known as natural philosophy: "Of the Stars," "Of the Air," "Of Minerals," "Of Insects"—all subjects pertaining to Mather's efforts to discover the mysteries of the natural world and, just as importantly to him, the one who created it.

This ultimately religious desire to understand led the most influential clergyman of his day to become an advocate of observation and experiment as means of increasing knowledge. "Let us proceed . . . and lift up our Eyes unto the Stars," Mather wrote. "The Telescope, invented the Beginning of the last century, and improved now to the Dimensions even of eight feet, whereby Objects of a mighty Distance are brought much nearer to us; is an Instrument wherewith our Good GOD has in a singular manner favoured and enriched us: A Messenger that has brought unto us, from very distant Regions, most wonderful Discoveries." Just as scripture can help us understand the world, Mather reasoned, so, too, can scientific instruments like the telescope help us see even the most distant elements of creation.

Mather's longing to be not just a man of God but also what would now be called a man of science put him in an interesting position when a series of smallpox epidemics devastated New England during the late seventeenth and early eighteenth centuries. The highly contagious virus came to port cities like Boston on incoming ships and by some accounts killed up to 25 percent of those infected. To a Puritan like Mather,

Cover of the pamphlet *Inoculation of the Small Pox as Practised in Boston* by William Douglass, 1722.

the cause of any catastrophe was usually clear enough. Years before, when he heard of an earthquake that destroyed the city of Port Royal in Jamaica, he believed the ground had shaken because the people of that colony had fallen into the "heathen" excitement of visiting fortune tellers. Mather likewise saw a connection between the number of miscarriages in Boston and the "great and visible decay of piety" in the city. Such hardships could be understood as expressions of God's wrath, he then believed, and so the best recourse Christians had was to atone for their sins.

Yet when sickness hit close to home, Mather began to rethink this position. With his children suddenly showing symptoms he had seen when visiting the dying in "venomous, contagious, loathsome Chambers," as he called the houses of the afflicted, he began to wonder: If it was in humanity's power to counteract illness through the God-given gift of the intellect, wouldn't it be wrong to fail to do so?

Eager to prove his learning equal to that of any of his peers back in England, Mather read as many European medical texts as he could lay his hands on. Emboldened by these interests, Mather proposed a radical new treatment be used to fight smallpox: inoculation. But doing so meant breaking ranks with both his country's Puritan elite and what then passed for common sense: If plagues were sent by God, to work against them was blasphemy.

The controversy that followed was both the first conflict between science and religion in America and the first media-addled public health disaster. Printer James Franklin (Benjamin's brother) fanned the flames by publishing screeds against inoculation generally and Mather personally. The pamphlet war between these two factions—for inoculation and against it—played out for months. Many attacks on Mather and his views focused on the source of his information. For an infamous witch hunter, he was surprisingly broad-minded about where he looked for new ideas. Mather had first learned of inoculation's efficacy from Boston's enslaved population, which had received the treatment before being kidnapped, sold, and brought to America. Specifically, Mather learned of inoculation from a man named Onesimus, who was enslaved by Mather himself. Long before he came across the procedure in his European medical texts, Mather wrote, "I had from a servant of my own an account of its being practised in Africa."

In a 1716 letter to the Royal Society of London, Mather recounted that Onesimus displayed a scar on his arm and explained that he had "undergone an Operation, which had given him something of ye Small-Pox, and would forever preserve him from it." Mather investigated among other Africans in Boston and found that inoculation was widespread. Enslaved men and women with scars on their arms were even sold for higher prices since they were known to be immune. Armed with this knowledge, Mather promoted inoculation as the best protection against the virus and urged Boston's doctors to adopt the practice. He convinced one of the colony's most preeminent physicians, Dr. Zabdiel Boylston, to administer the procedure during the summer of 1721.

Many Bostonians were skeptical of Mather's crusade for inoculation. Some, suspicious of African medicine, accused Mather of "Negroish" thinking. Others raised theological objections. Rev. John Williams quoted Matthew 9:12 ("It is not the healthy who need a doctor, but the sick.") as evidence that inoculation violated natural law. Williams also maintained that because inoculation could not be found in the Bible, it must not be the will of God. Many Puritans believed that God was showing them disfavor through the smallpox epidemic and therefore considered it a distraction to cure the disease rather than address their errant ways.

Portrait of Cotton Mather, 1728. Mather is best remembered as an influential Puritan minister, but he was also an amateur scientist and a tireless promoter of inoculation.

As Mather's opinions became known, anger at the practice of inoculation became more fervent and even violent. A lit grenade thrown through the minister's window came with a note: "Mather, you dog! I'll inoculate you with this!" While the fuse apparently fell out when it hit the floor, the bomb's intent was clear. These lines from a 1721 anti-inoculation pamphlet get to the heart of it: "Is it not taking God's work out of his hands?... Is it any better than dictating what measure of his judgement we intend to have?" In other words, if inoculation worked, then God was not in control, and if God was not in control, many feared they would not have their greatest source of solace just when they needed it most.

Fortunately, others offered support for this new medical procedure, and, interestingly, some of the most forceful and persuasive arguments in favor of inoculation were not scientific but theological. Prominent pastors, including Benjamin Colman and William Cooper, publicly defended inoculation experiments, arguing that the will of God could also be discerned in nature. Mather emphasized the importance of empirical observation and argued that inoculation adhered to Puritan principles; it was, for him, an example of the fruits borne of divinely given reason.

The controversy continued until the effectiveness of the procedure was demonstrated by both firsthand experiences and statistics presented by Boylston and Mather, which showed the smallpox mortality rate drop to 1 or 2 percent with inoculation treatment. As faith in the procedure grew, it became common practice to take one's children for inoculation between smallpox outbreaks. Over the course of the eighteenth century, inoculation transformed in the public imagination from a potential abomination to a gift that humans could use to save themselves.

A Pharmacist's Herb Garden

While New England was divided by questions of religion and medicine, elsewhere in early America their connection was more constructive. In the same decade that saw Cotton Mather debating inoculation and divine will, a group of Ursuline nuns established a hospital and school in New Orleans. One of the elders of the group, Sister Francis Xavier Hebert (ca. 1697–1762), soon planted an herb garden from which she compounded plants into medicines, making her one of the earliest pharmacologists in the French colony. Though for thirty years she served as chief administrator of the Royal Hospital, she never ceased the practical work that was her passion. Using the mortar and pestle shown here, Hebert helped launch the long Catholic tradition of health care in America.

Mortar and pestle used by Sister Francis Xavier Hebert, ca. 1697–1762.

Sister Francis Xavier was one of the first Roman Catholic nuns to journey from France to the Americas. Born Charlotte Hebert in Bayeux, Normandy, she joined the religious community that had been founded as the Company of Saint Ursula in 1596. It was a time of spiritual revival and social turmoil as battles raged between Catholics and Protestant Huguenots, which unexpectedly created an opening for the formation and growth of new Ursuline foundations. Dozens were founded in the first years of the community's existence.

By the time Hebert took her vows, the Ursuline Sisters were a well-established teaching order, operating more than three hundred schools offering girls instruction in reading, writing, and arithmetic. At the age of thirty-five Sister Francis Xavier traveled with a sister to the new French colony of Louisiana, where she joined a small group of nuns who had arrived two years before. They had been recruited by the Company of the Indies, which had need of a military hospital on the Gulf Coast. Sister Francis Xavier and the others agreed on the condition that they could also work toward improving education in the colony. Two of the institutions the sisters established—an Ursuline religious community and a school—both continue to this day. The convent constructed for their use was completed in 1745, making it the oldest extant building in the Mississippi Valley.

Soon after her arrival, Sister Francis Xavier began work on her herb garden. The poor conditions in the facilities initially made available to the Ursulines, along with the scarcity of medicines in the colony generally, made preparing their own treatments essential. Nearly all of these medicines began in the ground, and in Sister Francis Xavier's hands, they would soon cure and comfort members of the surrounding community. The herb garden became critically important to the lives of those within the convent and beyond.

Hebert's work compounding medicines at the Royal Hospital has given her the distinction of being called "the first woman pharmacist in the New World." As a history shared in the *Old Ursuline Convent Cookbook* notes, "The herbs that can be grown in the New Orleans climate successfully are thyme, sage, rosemary, mint, sweet marjoram, basil, lavender, anise, caraway, bene, sage, catnip, coriander, dill, fennel, horehound, pot marigold, dandelion, penny royal, rue, summer savory, tansy and tarragon." The sisters made use of all of these in the hospital ward, crafting the teas, infusions, tinctures, and distillates that were the only remedies available for tending to the injured and the ill.

Though occupied unrelentingly with the responsibilities of cultivation and administration, Sister Francis Xavier also became known for the attention she paid to the girls in the convent's care. The Ursulines educated young women of European, Native American, and African descent. Hebert, following the completion of her duties in the hospital, devoted herself to orphans. "She had a zeal and charity for these poor children," a contemporary account noted, "with whom she was constantly employed."

Given her concern for the well-being of others, it is perhaps surprising that maintaining her own health was apparently not regarded with similar urgency. In keeping with popular ascetic religious practices of the day, she was drawn to the mortification of the flesh. "She gave herself with zeal to all the austerities which a spirit of penance could invent," her obituary said. "She never took but one meal each day, and in that one meal she never ate that but what was most common and crude."

Sister Francis Xavier Hebert undeniably advanced medical practice in early America. She also helped establish a tradition of women's autonomy at a time when this was far from given. By the time the Louisiana Purchase made the heavily Catholic French territories part of the young United States in 1803, the spiritual descendants of Sister

The Landing of the Ursulines, August 7, 1727 **(detail) by Marie-Madeleine Hachard, 1859. The Ursulines were the first Roman Catholic nuns in what is now the United States.**

Francis Xavier had been managing their own affairs in New Orleans for more than seventy years. Concerned for their rights and their property in their new nation, the Ursulines wrote to President Thomas Jefferson to inquire if they had anything to fear. In his reply, Jefferson assured the sisters that their convent would not only be protected but remain "sacred and inviolate."

Even as the bounds of nations and governments shifted around them, the Ursuline community held firmly onto to its own authority. This was of course rooted largely in faith; thanks to Sister Francis Xavier Hebert, it was also rooted in science. In 2003 the World Health Organization and the National Association of Catholic Charities recognized her accomplishments in pharmaceuticals and philanthropy, which began with a simple mortar and pestle made of wood.

Electric Fire

Benjamin Franklin's lightning rod, ca. 1752.

In the eyes of many American colonists, a storm was a mysterious and dangerous phenomenon. Until the mid-eighteenth century, lightning was considered a supernatural force that could do strange things: it could poison wine and meat, or it could leave symbolic imprints on the flesh of those animals and humans unlucky enough to be struck. Stories circulated about lightning strikes engraving crosses or even Latin prayers onto human flesh. Such "lightning pictures" were also said to sometimes mirror the immediate surroundings at the time of the strike.

Many Christians believed that violent weather resulted from either God or a "diabolical agency." Lightning was particularly concerning because the tallest structure in most eighteenth century European and New England towns was the cathedral tower or church steeple. These structures were ideal lightning targets, and church bell ringers were often electrocuted. Christians therefore developed a variety of special prayers and procedures in the attempt to protect against lightning. Some believed that lightning could be dispelled by the ringing of church bells that had been engraved with Latin admonitions and baptized with water from the River Jordan. Discs of wax known as Agnus Deis, impressed with the figure of a lamb (representing the "lamb of God") and blessed by the pope, were considered to have some protective power as well. Other Christians were skeptical of such precautions. Martin Luther accepted that lightning was caused by demonic agency but doubted that anything so powerful would be frightened of church bells.

Enter Benjamin Franklin (1706–1790), Philadelphia's polymath Deist with a kite and a key. Franklin began researching the phenomenon of electricity in 1746, and in 1750 he published a proposal for an experiment to prove that lightning was electricity by flying a kite in a lightning storm. In 1752 a French physicist conducted Franklin's proposed experiment, using an iron rod instead of a kite, and successfully extracted electrical sparks from a cloud. Franklin described the experiment in the *Pennsylvania Gazette* in 1752 and mounted the first grounded lightning rods on the Pennsylvania State House and the Pennsylvania Academy. In 1753 he published instructions in his *Poor Richard's Almanack* for "How to Secure Houses, etc., from Lightning" using a grounded rod.

Franklin's lightning rod sparked immediate controversy. Among his ardent defenders was none other than his fellow Founding Father John Adams, who saw in the consternation of some a pattern of religious alarmism in the face of innovation like that aroused by inoculation.

"Surely the Thunder of Heaven is no more supernatural than the Rain, Hail or Sunshine of Heaven."

Franklin's most vocal critic, meanwhile, was the French priest and physicist Jean-Antoine Nollet (1700–1770). Nollet was skeptical of the lightning rod on both scientific and religious grounds. As a scientist, he was dubious that the overwhelming power of a thunderstorm could be diverted by a thin stick of iron. As a priest, Nollet considered even attempting it to be prideful and "impious." This man-made rod was no match for the "chastening rod" of God.

Franklin had little patience for Nollet's objections. During much of his life Franklin counted himself an adherent of Deism, which held that there is a supreme being who created the world but that this God generally left humans alone. Franklin believed that the nature of the divine was essentially unknowable to human minds, but that we can deduce that God cares at least somewhat for humans to the extent that humanity had been endowed with reason. This is reflected in the language Franklin used to introduce the lightning rod in *Poor Richard's Almanack*: "It has pleased God in His goodness to mankind at length to discover to them the means of securing their habitations and other buildings from mischief by thunder and lightning."

Franklin balked at Nollet's theological argument that diverting lightning would somehow be impious. He claimed that Nollet "speaks as if he thought it Presumption in man to propose guarding himself against the Thunders of Heaven!" "Surely," Franklin continued, "the Thunder of Heaven is no more supernatural than the Rain, Hail or Sunshine of Heaven, against the Inconvenience of which we guard by Roofs & Shades without Scruple."

As further refutation of Nollet's objections, Franklin outlined an additional discovery: that "it is not Lightning from the Clouds that strikes the Earth, but Lightning from the Earth that strikes the Clouds." Using a device that stores an electrical charge called a Leyden jar, Franklin had determined that thunderclouds are most often charged negatively. This meant that the force that created lightning originated in the Earth itself.

Franklin's lightning rod faced a second round of controversy in 1755, after an earthquake struck southern New England during the night, when many were asleep. The event inspired a fierce debate about God's will, electricity, and the causes of earthquakes. Although mid-eighteenth-century New Englanders had become increasingly doubtful that God intervened in the natural world for the purposes of retribution, many Christians worried that God was shaking the Earth as punishment for human sins.

The Harvard-educated pastor Thomas Prince, a follower of the Massachusetts Puritan minister Cotton Mather, responded to the 1755 earthquake by reprinting a sermon he had written after the earthquake of 1727 titled "Earthquakes the Works of God, and Tokens of His Just Displeasure." In the reprint he added the opinion that earthquakes were caused by electricity, positing that Bostonians may have exacerbated the tremor by putting up so many lightning rods, which channeled electricity into the ground.

Portrait of Benjamin Franklin, 1879. The polymath scientist and diplomat sparked a theological controversy with his famous kite experiment.

Prince's publication prompted a feud with Harvard professor John Winthrop, great-great-grandson of Governor John Winthrop, the founder of the Massachusetts Bay Colony who famously declared America a biblical "city upon a hill." Professor Winthrop proposed an alternative account of earthquakes, having traced the pattern of fallen brick structures to show that the Earth had undulated. Winthrop also criticized Prince's notion "that it is possible, by the help of a few yards of wire, to 'get out of the mighty hand of God.'" Winthrop believed that it was God's plan for humanity to discover the workings of the cosmos and that God would approve of humanity's efforts to protect themselves from lightning.

With this sentiment Benjamin Franklin wholeheartedly agreed. Like Cotton Mather's support of inoculation, Franklin's experiments with electricity raised what would prove to be a perennial American theological concern: that new technologies and scientific discoveries might subvert the moral order of the universe. Simultaneously, however, Franklin's religious justifications of his efforts also suggest that he viewed experiment and discovery as part of a divine plan.

Thomas Paine's Clockwork Universe

While his 1776 pamphlet *Common Sense* is often hailed as being as important to the revolutionary cause as the leadership of George Washington, Thomas Paine (1737–1809) achieved his most radical work later. *The Age of Reason*, written in part in a Paris prison, outlined Paine's religious and scientific philosophies, which for him were inseparable. Paine was among the most forceful advocates of Deism, a religious perspective that looked for divinity in nature's laws and had no use for the revelations of scripture. As *The Age of Reason* explained, Paine had come to his break with orthodox Christianity through natural philosophy, and particularly through the view of the vastness of the universe that astronomy provided him.

Paine learned astronomy initially through a book written by the Scottish astronomer and instrument maker James Ferguson, who also designed terrestrial globes providing three-dimensional views of the Earth. It was through the study of such instruments that Paine began to develop a religious worldview distinct from that offered by mainstream Christianity. Paine was particularly moved by a type of device known as an orrery, "a machinery of clock-work, representing the universe in miniature," that demonstrates the relative positions, sizes, and orbits of the planets and moons "as they really exist in what we call the heavens."

Orreries date back to the ancient Greeks, but they made a comeback during the eighteenth century. Advances in clock making and astronomy had made possible a new generation of machines that were remarkably precise, ornate, and complex. Less than two centuries after Galileo was tried for his support of heliocentrism, it had become commonplace in Atlantic intellectual circles to see the sun at the center of these mechanical maps of celestial bodies.

The Willard Orrery, on pages 42 and 43, was made by Boston clockmaker Aaron Willard Jr. Constructed approximately ten years after

A Philosopher Lecturing on the Orrery **by Joseph Wright of Derby, ca. 1766.**

Succeeding pages: Orrery made by Aaron Willard Jr., ca. 1820. When its crank was turned, this educational device replicated the revolutions of the planets and moons in our solar system.

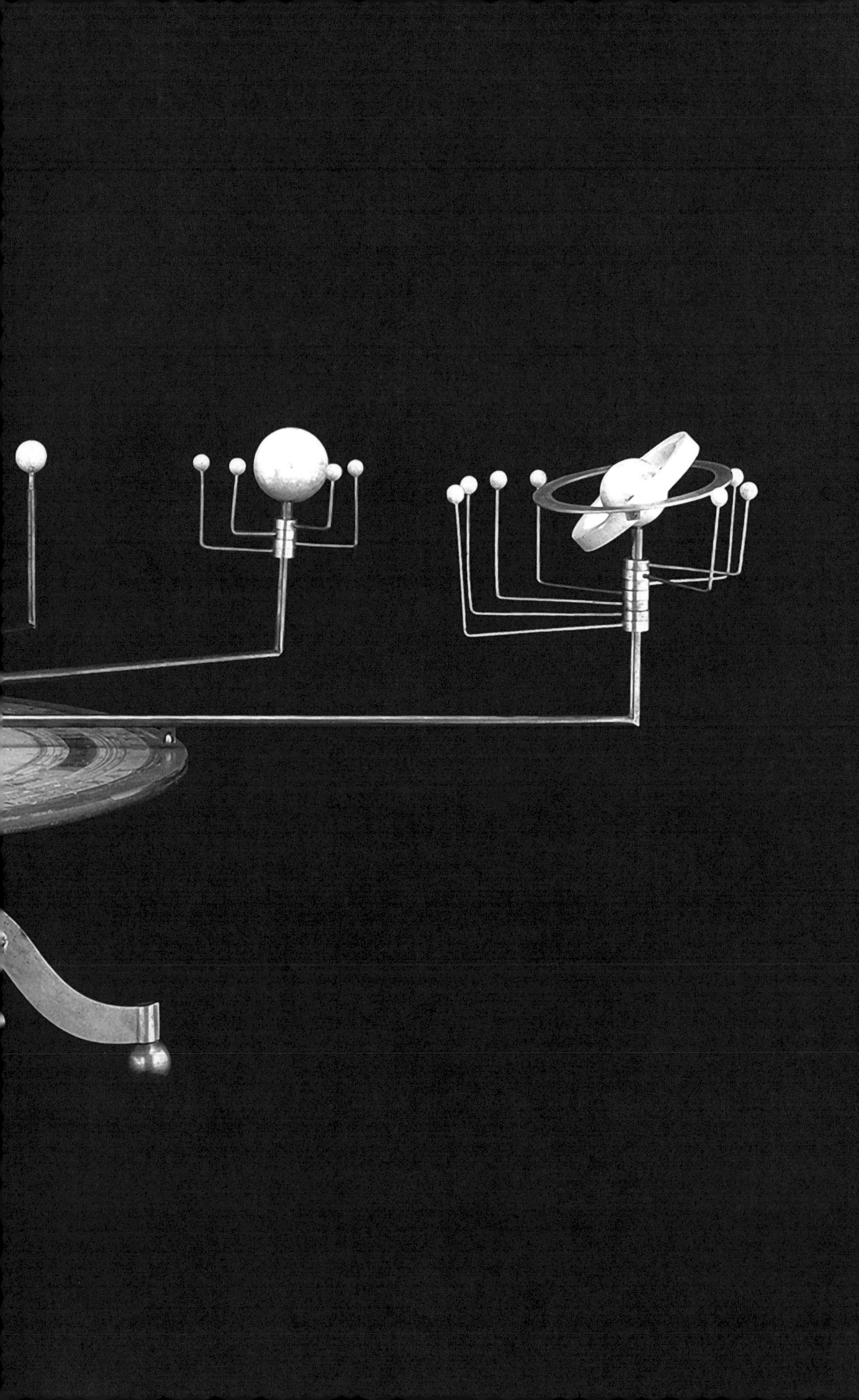

Paine's death, it was not the orrery that inspired Paine, but it is a vivid and representative example of the technology that forever changed him. Like many of the orreries produced during this era, the Willard Orrery uses clockwork mechanisms to precisely replicate motions of the Sun, the Earth and its moon, and the planets out to Saturn.

Not only technologically advanced, orreries were often constructed out of opulent materials. The Willard Orrery includes a Sun made of brass, planets made of ivory, and a mahogany horizon circle supported by four elegant brass legs. These instruments were meant to amaze; they were both literally and self-consciously revolutionary. In the context of the eighteenth century, orreries were akin to scientific miracles, built to inspire the kind of awe and wonder that might typically be associated with religion. The ca. 1766 painting *A Philosopher Lecturing on the Orrery* shown at the opening of this essay depicts this wonder in the faces of two children watching an orrery demonstration.

As Paine tells it in *The Age of Reason*, it was the orrery that led him to ultimately doubt Christian doctrine. Mastering the orrery, he wrote, helped him conceive of "the infinity of space, and the eternal divisibility

Engraving of an orrery, 1749. With their promise to explain "the Theory of the Heavens and the Earth," as this engraving notes, orreries inevitably challenged traditional religious understandings of the cosmos.

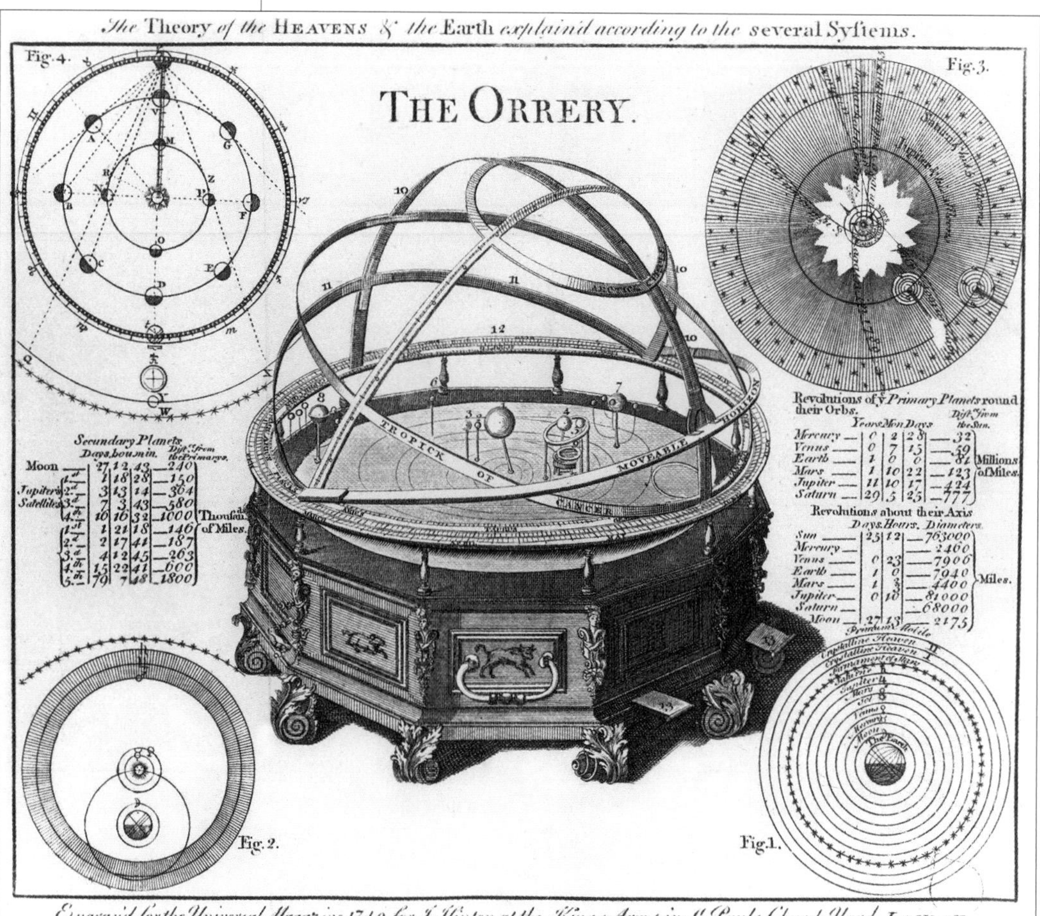

Portrait of Thomas Paine, 1793. Paine was an influential figure in the American Revolution. He was also deeply critical of religious institutions.

of matter." After comparing "the eternal evidence those things afford" with "the Christian system of faith," Paine found the latter wanting.

Paine's autobiographical story reflected an increasingly common belief during the revolutionary period: that, as science continued to map out the laws of the natural world, there would be less and less room for religion. In place of traditional scripture, Paine preached from what he referred to as "the Bible of Creation," by which he meant the world and the universe as observed. "The Bible of Creation is inexhaustible," Paine wrote. "Every part of science, whether connected with the geometry of the universe, with the systems of animal and vegetable life, or with the properties of inanimate matter, is a text as well for devotion as for philosophy, for gratitude as for human improvement."

"Every part of science is a text as well for devotion as for philosophy, for gratitude as for human improvement."

If Paine saw in the orrery evidence of religion's fallibility, he also saw a glimmer of something magnificent: an all-encompassing system of knowledge that might one day finally deliver on the theological promise of omniscience. What Paine envisioned would later be known as "consilience," the dream of a unified system of all knowledge. (Some of Paine's contemporaries also saw wonder in the orrery, albeit for different reasons. When teaching with the orrery, instructors at Harvard College were required to note that it demonstrated the genius of divine creation.)

Paine's critique of Christianity proved too much for many Americans. Late in life he did not enjoy the influence he'd had during the Revolution, though many future generations would celebrate his heterodox views about science's intellectual authority. For his part, Paine was optimistic about the future of religion and science, though he believed the former would transform into the latter. If "a revolution in the system of religion takes place, every preacher ought to be a philosopher," he said in 1794, "and every house of devotion a school of science."

Benjamin Bannaker's

PENNSYLVANIA, DELAWARE, MARY-
LAND, AND VIRGINIA

ALMANAC,

FOR THE

YEAR of our LORD 1795;

Being the Third after Leap-Year.

BANNAKER.

—PRINTED FOR—

And Sold by JOHN FISHER, *Stationer.*

BALTIMORE.

An Almanac of Strange Dreams

Almanacs were the eighteenth century's internet. The regularly printed miscellanies offered entertainment, inspiration, practical information, and a direct line (or at least as direct as one could get at the time) to the wider world. Almanacs were among the colonies' earliest printed matter. Relatively cheap and easy to acquire, they were found in homes more often than the Bible.

Yet to draw a distinction between the two—almanacs and Bibles—is not entirely accurate. After all, the two genres of literature informed each other. Just as the Bible is full of weights and measures along with sacred history, an almanac might include snippets of scripture alongside descriptions of the phases of the moon. Almanacs regularly offered weather forecasts, tide tables, and charts illustrating lunar and solar eclipses, all sharing pages with poems and proverbs that shaped the common culture of the day. They were a mirror of the varied interests of Americans as they left their colonial identities behind and struggled to determine what it might mean to be a new nation.

One of the most intriguing almanac creators in the early United States was the Black farmer, mathematician, and inventor Benjamin Banneker. Born in 1731 in Ellicott's Mills, Maryland, he became—through an inquiring mind and self-taught efforts—one of the first African Americans to gain distinction in science. His interest in religion overlapped with his scientific accomplishments, and together they profoundly affected the perception of African Americans, both enslaved and free.

A well-known figure in his day, Banneker frequently corresponded with the leaders of the new nation, to whom he strove to demonstrate the error of their ideas about racial limitations. Banneker believed that his almanacs, which precisely predicted eclipses and planetary alignments, could debunk the widespread racist notion that persons of African descent were intellectually inferior. His most notable exchange

Cover of *Benjamin Banneker's Pennsylvania, Delaware, Maryland, and Virginia Almanac*, 1795.

was with Thomas Jefferson, whose claims about Black inferiority Banneker publicly challenged.

On August 17, 1791, Banneker wrote to Jefferson, who was then secretary of state, disputing his assertion that "the blacks . . . are inferior to the whites in the endowments both of body of mind." In his letter, Banneker appealed to Jefferson to share his faith "that one universal Father hath given Being to us all, and that he hath not only made us all of one flesh, but that he hath also without partiality, afforded us all the Same faculties, and that however variable we may be in Society or religion, however diversified in Situation or colour, we are all of the same family, and Stand in the Same relation to him." To bolster his argument he also included the words of Phillis Wheatley, the famed enslaved poet of New England, along with abolitionist voices further making the case for equality.

Banneker also enclosed a handwritten manuscript of his forthcoming 1792 almanac. "Having long had unbounded desires to become acquainted with the Secrets of nature," Banneker wrote, "I have had to gratify my curiosity herein thro my own assiduous application to Astronomical Study, in which I need not to recount to you the many difficulties and disadvantages which I have had to encounter."

Jefferson replied almost immediately with praise for Banneker's almanac, calling it "a document to which your whole colour had a right for their justification against the doubts which have been entertained of them. . . . No body wishes more than I do to see such proofs as you exhibit that nature has given to our black brethren, talents equal to those of the other colours of men, and that the appearance of a want of them is owing merely to the degraded condition of their existence both in Africa and America."

Banneker later published his letter, along with Jefferson's response, in the 1793 edition of his almanac. Banneker was the first Black man to directly and publicly challenge Jefferson's prejudiced assumptions, and the exchange was circulated widely by abolitionists in the United States and Great Britain.

The cover of Banneker's 1795 *Almanac* opening this chapter features a woodcut portrait depicting the author dressed in the simple Quaker garb for which he was known. Though often associated with the group known as the Religious Society of Friends, his religious interests were in fact far more eclectic, as can be seen in his astronomical journal—which, as something of a personal almanac, blended his many interests on nearly every page. A single day's entries might include projections for eclipses, a list of church feast days, and a vivid account of his dreams.

In recent years it has been these dreams that have most excited new interest in Banneker as a figure in whom the complicated interplay of science and religion in early American can be seen. Some of his dreams focused on a mysterious shape known as Quincunx, which is at once an ancient Roman unit of measure, an astrological term referring to the precise position of the planets, and a symbol used in Senegalese Islam. While the journal never mentions its author having any known Muslims in his lineage, scholars at the Maryland Historical Society have hypothesized that such references may reflect knowledge passed down through the generations, perhaps beginning with Banneker's West African grandfather. Other dreams, meanwhile, seem to draw upon the rich spiritual and folkloric traditions of enslaved communities:

Postage stamp commemorating Benjamin Banneker, 1980. Banneker was a surveyor, clockmaker, and astronomer who correctly predicted eclipses and planetary alignments.

December 13, 1797

I Dreamed I saw some thing passing by my door to and fro, and when I attempted to go to the door, it would vanish. . . . At length I let in the infernal Spirit. . . . I know not what became of him but he was an ill formed being—Some part of him in Shape of a man, but hairy as a beast . . . but while I held him in the fire he said something respecting he was able to stand it, but I forget his words.

April 24, 1802

I dreamed I had a fawn or young deer; whose hair was white and like unto lamb's wool, and all parts about it beautiful to behold. Then I said to myself I will set this little captive at liberty, but I will first clip the tips of his ear that I may know him if I should see him again. Then taking a pair of shears and cutting off the tip of one ear, and he cried. . . . I did not attempt to cut the other but was very sorry for that I had done . . . and he ran a considerable distance then he stopped and he looked back at me . . . and he came and met me and I took a lock of wool from my garment and wiped the blood of [the] wound which I had made on him (which sorely affected me) I took him in my arms and brought him home and hold him on my knees.

"What are these dreams exactly?" the Maryland Historical Society has asked. "Are they riddles? Allegorical? Or can they be characterized as curiosities and nothing more?" While their meaning remains elusive, Banneker's practice of recording his strange and intriguing dreams alongside his detailed astronomical observations is suggestive of the ways in which spiritual notions and scientific ideas have long shared space in the American psyche.

Religious Freedom and the Air We Breathe

Glass matrass used by Joseph Priestley, ca. 1790.

Joseph Priestley (1733–1804) wore many hats. The famed chemist is best known for discovering oxygen, but he was also a minister, a natural philosopher, and a groundbreaking political theorist. During his lifetime Priestley also helped establish one of America's most cherished traditions: religious tolerance.

Although he was born in England, Priestley spent the last ten years of his life in Pennsylvania. Like many immigrants before him, Priestley came to the United States seeking refuge from religious persecution, after a mob burned down his English home in 1791. ("Antagonists think they have quenched his opinions by sending him to America," Thomas Jefferson wrote of the events, "just as the Pope imagined when he shut up Galileo in prison that he had compelled the world to stand still.")

Priestley's most famous discovery can be traced to 1771, when he began a series of experiments to distinguish between the different "airs" that he believed to be mixed in our common atmosphere. Priestley used airtight glass vessels, like the bell jar pictured on page 52 from his Pennsylvania laboratory, to isolate various substances and measure their interactions within an enclosed space. Priestley observed, for example, that a candle would burn out when deprived of free-flowing air. Similarly, a mouse or spider eventually died when trapped in an airtight space, and they died even faster if their chambers contained a burning flame. Yet, to Priestley's surprise, a sprig of mint could survive in those same conditions. Not only that, plants seemed to restore something to the air that the flame and living creatures had exhausted.

Priestley kept burning things for science. He knew that when iron rusted it counterintuitively gained weight, forming a "calx," or what is now known as an oxide. He sealed a mouse in a glass jar with a sample of mercury calx and then ignited the calx using a burning glass, which concentrates the sun's rays with a convex lens. A mouse in an airtight space lived longer, Priestley determined, when trapped with burning calx. Priestley deduced that the calx must be releasing something good back into the contained atmosphere. All that was left was to try this miraculous new air for himself. Priestley burned some mercury calx, took in a deep breath, and reported that his "breast felt peculiarly light and easy for some time afterwards." Priestley had discovered oxygen.

"Dephlogisticated air," as he called it, was just one of many important discoveries that Priestley made during his lifetime. His research on electricity earned him praise from Benjamin Franklin, who nominated Priestley to join the prestigious Royal Society of London. Priestley is also credited with the invention of carbonated water, which he created by standing over a brewery vat and pouring water back and forth between two glasses. (He tested this on himself first, too, later calling it his "happiest" discovery.)

The intrepid scientist and inventor was also bubbling with religious ideas. Priestley was a Christian minister whose liberal, rationalist theology was considered brilliant and audacious by many of his contemporaries. Combining Enlightenment values with Christian theism, Priestley outlined a metaphysics that was equal parts science and religion. He believed that the revelations of scripture and natural law

Priestly believed that the revelations of scripture and natural law must coincide harmoniously and that Christianity must therefore be purged of distorting myths and errors.

Bell jar used by Joseph Priestly, ca. 1790. After fleeing religious persecution in England, the chemist and minister continued his scientific work in Northumberland, Pennsylvania, where he used this bell jar in his laboratory.

***Joseph Priestley* by Albert Rosenthal after Gilbert Stuart, 1921.**

must coincide harmoniously and that Christianity must therefore be purged of the distorting myths and errors that had accumulated over the centuries. This meant doing away with any religious precepts that failed to meet scientific muster, including the Trinity (which Priestley dismissed as supernatural), the divinity of Jesus (a concoction, he argued, of fifth-century theologians), and the Calvinist doctrine of election (which he considered logically irreconcilable with the notion of an impartial loving God).

As he reflected later in his autobiography, Priestley had a habit of embracing whatever was "generally called the heterodox side of almost every question." He left no stone unturned, even going so far as to question the axiom that there was an immaterial soul that existed separately from the body. In his *Disquisitions Relating to Matter and Spirit*, for example, Priestley argued, first, that Descartes was demonstrably wrong when he split the mind and body into separate essences, and, second, that the Anglicans of his day were anachronistically imposing mind/body dualism onto Biblical texts, despite evidence that no such distinction existed amongst the Jews and early Christians of the ancient world. Priestley did not consider such criticism to be blasphemous but quite the opposite: to rid Christianity of such errors, he thought, would only bring humanity closer to perfect knowledge of God and the universe.

One of Priestley's central theological concerns was theodicy, or the "problem of evil." How is it possible to reconcile the existence of evil and suffering in the world with the notion of an omnipotent, omnibenevolent God? If God is all-good and all-powerful, in other words, why is there suffering on Earth? Theodicy is a concern as old and variegated as monotheism itself, but it was taken up anew by Priestley's generation during the chaotic and revolutionary eighteenth century.

Priestley put science at the center of his theodicy. In his 1768 *An Essay on the First Principles of Government*, Priestley argued that, rather than creating a perfect world from the start, God created the conditions through which humans could make progress themselves. Science, he argued, is one of the best engines for this progress: it eliminates superstition, promotes human welfare, subverts arbitrary political power, and leads the human mind out of "labyrinths of error." Thus, for Priestley, evil in the world was a temporary setback in humanity's ordained journey toward perfection.

Priestley's theodicy, in turn, made him a particularly open scientist. During a time when British and French spies were infiltrating scientific laboratories, Priestley was an open book. He shared his discoveries publicly and immediately considered how they might be applied to medical uses that would ameliorate human suffering. Priestley was also open to intellectual criticism, which he considered crucial to human progress. In the preface to *Experiments and Observations on Different Kinds of Air and Other Branches of Natural Philosophy*, he wrote, "I find it

Lithograph by Charles Joseph Hullmandel, after a painting by Johann Eckstein, 1791. The destruction of Priestley's home in Birmingham, England, became a rallying cry for those who fought against religious persecution, including Thomas Jefferson.

absolutely impossible to produce a work on this subject that shall be anything like *complete*. Every publication I have frankly acknowledged to be very imperfect.... Paradoxical as it may seem, this will ever be the case in the progress of natural science, so long as the works of God are, like himself, infinite and inexhaustible. In completing one discovery we never fail to get an imperfect knowledge of others, of which we could have no idea before; so that we cannot solve one doubt without creating several new ones."

Resistance to Priestley's ideas bubbled up in his home country of England. Priestley was part of a religious minority known as the "Dissenters," a term for English Protestants who did not conform to the doctrines and practices established by the Church of England. Dissenters were excluded from full participation in the Anglican state. They were barred from holding large-scale religious gatherings and constrained in their ability to train new ministers. Dissenters would not be employed in positions of state power unless they publicly received communion from the Church of England.

Priestley was an outspoken advocate of the Dissenters and a critic of religious intolerance everywhere. To further this cause, Priestley put society itself in the bell jar. He wrote sweeping analyses of social and religious history seeking to prove that intellectual progress only accelerates when more religious opinions are allowed to compete and thrive in the public square. Priestley's political philosophy reflected his scientific theodicy. The truth will out, he argued, and happiness will increase if governments stop intermeddling in religious affairs.

Many English Dissenters, including Priestley, were critical of British colonialism and openly supportive of the French and American Revolutions. Tensions erupted in July 1791 when a group of rioters burned down thirty Dissenter homes and churches, including Priestley's. With his belongings destroyed and his laboratory reduced to ashes, Priestley fled with his family to London and then Pennsylvania, where he would settle for the remaining years of his life.

The United States would prove much more welcoming to Priestley's ideas. Newspapers rejoiced at his arrival, political parties jostled for his support, and the University of Pennsylvania offered him a position teaching chemistry. Priestley turned down the professorship, but he couldn't resist an invitation to give a series of sermons on Unitarianism in Philadelphia, which helped lead to the founding of the first Unitarian church in America. As a house of worship built on the reputation of a scientist, it is a fitting illustration of the widespread assumptions of the compatibility of religion and science during this era, though these assumptions were soon to change.

EVO
LU
TIONS

Midway through the nineteenth century, Charles Darwin's theory of evolution caused a sea change in the relationship between religion and science, but it was not the only scientific advancement that had dramatic effects on the beliefs of Americans.

The age of Darwin was a period of dizzying growth in scientific understandings of life itself, drawing on data gleaned through the study of fossils, geology, comparative anatomy, and the global examination of life-forms from the simple to the complex. Technological innovations like the telegraph, electricity, and photography had surprising spiritual repercussions, while improved astronomical devices provided better understandings of the universe, which both affirmed a sense of wonder at the cosmos and further challenged explanations of its creation. All the while, questions concerning the perceived boundaries separating religion and science—Could either provide insight into the other? Who determined the limitations of each?—gave rise to self-taught practitioners who used unorthodox means, including quilting and other arts, to expand understandings of what it meant to contribute to spiritual or experimental knowledge.

Even as the notion that religion and science are locked in perpetual battle gained ground, faith traditions adapted to maintain a sense of compatibility with science as an ascendant cultural force. The leading boosters of theories of inevitable conflict, meanwhile, revealed themselves to be not without religious biases of their own. During this tumultuous and inventive century, points of intersection multiplied, pointing toward further moments of reckoning ahead.

"Solar System" quilt (detail) by Ellen Harding Baker, 1876. Baker made this quilt to illustrate her astronomy lectures in churches and community halls across Iowa.

The
Life and Morals
of
Jesus of Nazareth
Extracted textually
from the Gospels
in
Greek, Latin
French & English.

The Bible Surgeon

The Life and Morals of Jesus of Nazareth **by Thomas Jefferson, ca. 1820.**

For all his influence on science and theology, Joseph Priestley's greatest impact on American society may have been through his friendship with Thomas Jefferson (1743–1826). While the latter was an Anglican by baptism, his spiritual opinions late in life were singularly influenced by the Unitarian chemist, who introduced the third president to a reason-driven approach to religion that eventually would yield the experimental volume of scripture that has come to be known as the Jefferson Bible.

Priestley and Jefferson met in person for the first time in Philadelphia in 1797 and then corresponded regularly on topics ranging from the life of Jesus to the failures of the Adams administration. Jefferson purportedly gifted Priestley with a wooden cane that was donated to the Smithsonian in 1888. After his death, Priestley was remembered as an "excellent pedestrian" with a preference for such long, simple walking sticks.

Even before their first meeting, Jefferson had been profoundly affected by Priestley's writing. Jefferson had read *History of the Corruptions of Christianity* (1782), in which Priestley aimed to deconstruct Christian orthodoxy, historicize the Bible's supernatural claims, and restore the original message of Jesus. Jefferson had himself been struggling to reconcile Christianity with natural philosophy, and in Priestley's writings he finally found an agreeable synthesis. In a letter to John Adams, Jefferson wrote that he had read Priestley's pages "over and over again; and I rest on them . . . as the basis of my own faith."

Priestley's impact on Jefferson would, in turn, help shape some of the founding principles of the United States. Historians argue that the Declaration of Independence was likely influenced by Priestley's writings on civil government and religious tolerance. And he was indirectly responsible for the creation of perhaps the last monumental work of

Jefferson's monumental life, the Jefferson Bible. A labor of love during his long retirement, the eighty-four-page redacted edition of the New Testament, which Jefferson called *The Life and Morals of Jesus of Nazareth*, was the product of decades of thinking about scripture, religion, and the latter's role in society. He had first asked his learned friend Priestley to undertake the work of paring down the Gospels to their essential elements, using the lens of reason as guide. When Priestley demurred, Jefferson took up the challenge himself.

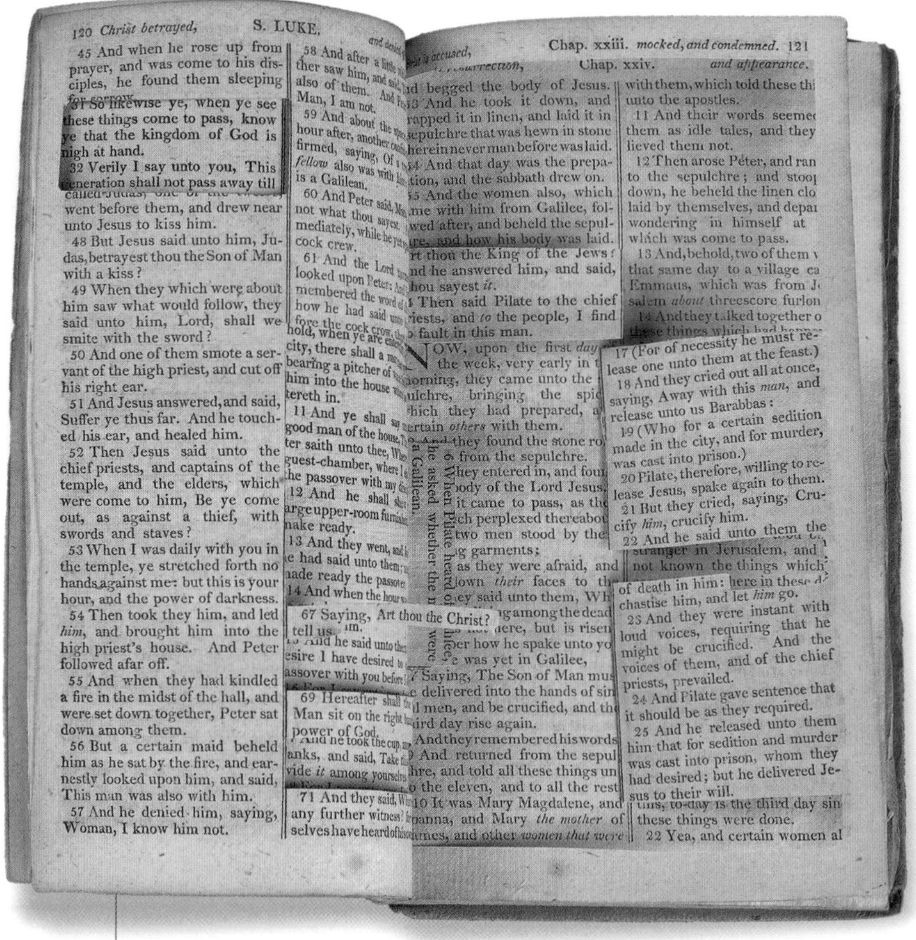

A New Testament from which Jefferson cut carefully chosen passages (left), and the book he made with the selected extracts (right). To compose *The Life and Morals of Jesus of Nazareth*, Jefferson drew upon multiple copies of the Gospels, using a penknife to separate sections containing supernatural elements from those he deemed reasonable. Only the latter made it into his redacted scripture.

This do-it-yourself spirit was, of course, entirely in keeping with his character. Jefferson had little doubt that, if anyone were up to the challenge of editing the Good Book, it would be him. As the early Jefferson biographer William Eleroy Curtis noted,

No problem was too abstruse for him to grasp. He seldom asked advice or assistance from others. . . . There is scarcely a subject in the entire range of human inquiry upon which Jefferson did not express his views in writing with fearlessness with absolute faith in his own convictions and judgment. He discusses art, architecture, the treatment of infants,

meteorology, music, astronomy, the practice of medicine, the breeding of sheep, the science of government, the apparel of women, the origin of meteoric storms, and the temperature of the moon as freely as politics or religion. In all the sciences he advanced propositions and solved problems with equal audacity. In criticism he was caustic and reckless and commends with the same freedom that he condemns. He rejects the Mosaic account of the creation and the flood as fiction and pronounces the Gospels the sublimest code of morals ever conceived.

Though such biblical opinions were firmly held by Jefferson during his time in the White House and resulted in what might be called a rough draft of the Jefferson Bible in 1804, he did not turn in earnest to the project he had proposed to Priestley until his retirement at Monticello.

Beginning in 1819, the seventy-seven-year-old Jefferson worked with a penknife and glue to excise sections from the Gospels in four languages—English, French, Greek, and Latin—and paste them onto the pages of a blank book. Anything that could not be accepted by Jefferson's scientific mind was left on the cutting-room floor.

6

CAPUT V.

CHAPITRE V.

Sermon sur la Montagne.

Mt. 5

AND seeing the multitudes, he went up into a mountain: and when he was set, his disciples came unto him:
2 And he opened his mouth, and taught them, saying,
3 Blessed *are* the poor in spirit: for their's is the kingdom of heaven.
4 Blessed *are* they that mourn: for they shall be comforted.
5 Blessed *are* the meek: for they shall inherit the earth.
6 Blessed *are* they which do hunger and thirst after righteousness: for they shall be filled.
7 Blessed *are* the merciful: for they shall obtain mercy.
8 Blessed *are* the pure in heart: for they shall see God.
9 Blessed *are* the peace-makers: for they shall be called the children of God.
10 Blessed *are* they which are persecuted for righteousness' sake: for their's is the kingdom of heaven.
11 Blessed are ye when *men* shall revile you, and persecute *you*, and shall say all manner of evil against you falsely, for my sake.
12 Rejoice, and be exceeding glad; for great *is* your reward in heaven: for so persecuted they the prophets which were before you.

L. 6

24 But woe unto you that are rich! for ye have received your consolation.
25 Woe unto you that are full! for ye shall hunger. Woe unto you that laugh now! for ye shall mourn and weep.
26 Woe unto you when all men shall speak well of you! for so did their fathers to the false prophets.

Mt. 5

13 Ye are the salt of the earth: but if the salt have lost his savour, wherewith shall it be salted? it is thenceforth good for nothing, but to be cast out, and to be trodden under foot of men.
14 Ye are the light of the world. A city that is set on an hill cannot be hid.

Thomas Jefferson **by Rembrandt Peale, 1805. The third president of the United States spent much of his life trying to reconcile science and religion.**

"Jesus did not mean to impose himself on mankind as the son of God," Jefferson wrote. The miracles and supernatural elements that make up so much of the traditional story of Jesus's life Jefferson regarded as fictions grafted on to the biography of a historically significant figure. The dogmas that had grown out of these fictions, he believed, were little more than the "abracadabra" of priests. To the sage of Monticello, the man from Nazareth was a great teacher and moral exemplar, and that was enough.

Frequently accused of heresy and even atheism, Jefferson knew that his cutting approach to scriptural interpretation would be too much to accept for those with more traditional perspectives. He discussed *The Life and Morals of Jesus Christ* only with a few friends and never imagined it would be widely distributed.

Along with Priestley, another of Jefferson's most trusted confidants on matters of faith was the patriot and physician Benjamin Rush (1746–1813). Long before he put penknife to paper to craft his alternate scripture, Jefferson wrote the doctor a letter outlining his thoughts on Christianity, as well as his frustration that they were so often misperceived. His views were, he wrote, "the result of a life of inquiry &

reflection, and very different from that anti-Christian system imputed to me by those who know nothing of my opinions. To the corruptions of Christianity I am indeed opposed; but not to the genuine precepts of Jesus himself. I am a Christian, in the only sense he wished any one to be; sincerely attached to his doctrines, in preference to all others; ascribing to himself every human excellence; & believing he never claimed any other."

While Jefferson was reluctant to broadcast his personal beliefs, his thoughts on religion in general were well known. The one book he published in his lifetime, *Notes on the State of Virginia*, includes a section on religion that captures the purpose and possibilities of matters of the spirit as he understood them. In Jefferson's view, the various religions of the world inevitably interact in ways at once contentious and transformative. "Uniformity of opinion," he said, was neither desirable nor attainable. Assessing the function of religious diversity in his young country, he noted that each sect was a moral censor of the others around it. This was not a call for mere tolerance; it was a recognition that faiths exist not in a vacuum but in relation to one another.

Elsewhere Jefferson said he was "of a sect by myself." Yet the reception of *The Life and Morals of Jesus of Nazareth* suggests that many Americans shared his views. Having produced a single copy of the manuscript in his lifetime, he bequeathed the leather-bound volume to his daughter Martha in his will. It passed down through generations of his descendants until his great-granddaughter Carolina Randolph sold it to the Smithsonian's chief librarian Cyrus Adler in 1895. Nine years later, an act of Congress decreed that the text be published for the first time. For fifty years, every newly elected senator received a copy of the Jefferson Bible upon taking the oath of office.

It was fitting that the manuscript was passed down through family, as Jefferson had always made his most personal statements regarding religion to those closest to him. In a letter written in 1787, he had urged a nephew of his to consider all sides when it came to questions of religion. If, Jefferson wrote, you hold the belief "that there is no God, you will find incitements to virtue in the comfort and pleasantness you feel in its exercise, and the love of others which it will procure you. If you find reason to believe there is a God, a consciousness that you are acting under His eye, and that He approves you, will be a vast additional incitement.... Your own reason is the only oracle given you by heaven."

The Life and Morals of Jesus of Nazareth is clear evidence that Jefferson not only gave this advice but lived by it. Scripture in his hands became a laboratory in which to test hypotheses about religious truth. Two hundred years later, Americans continue to study the results of his experiment.

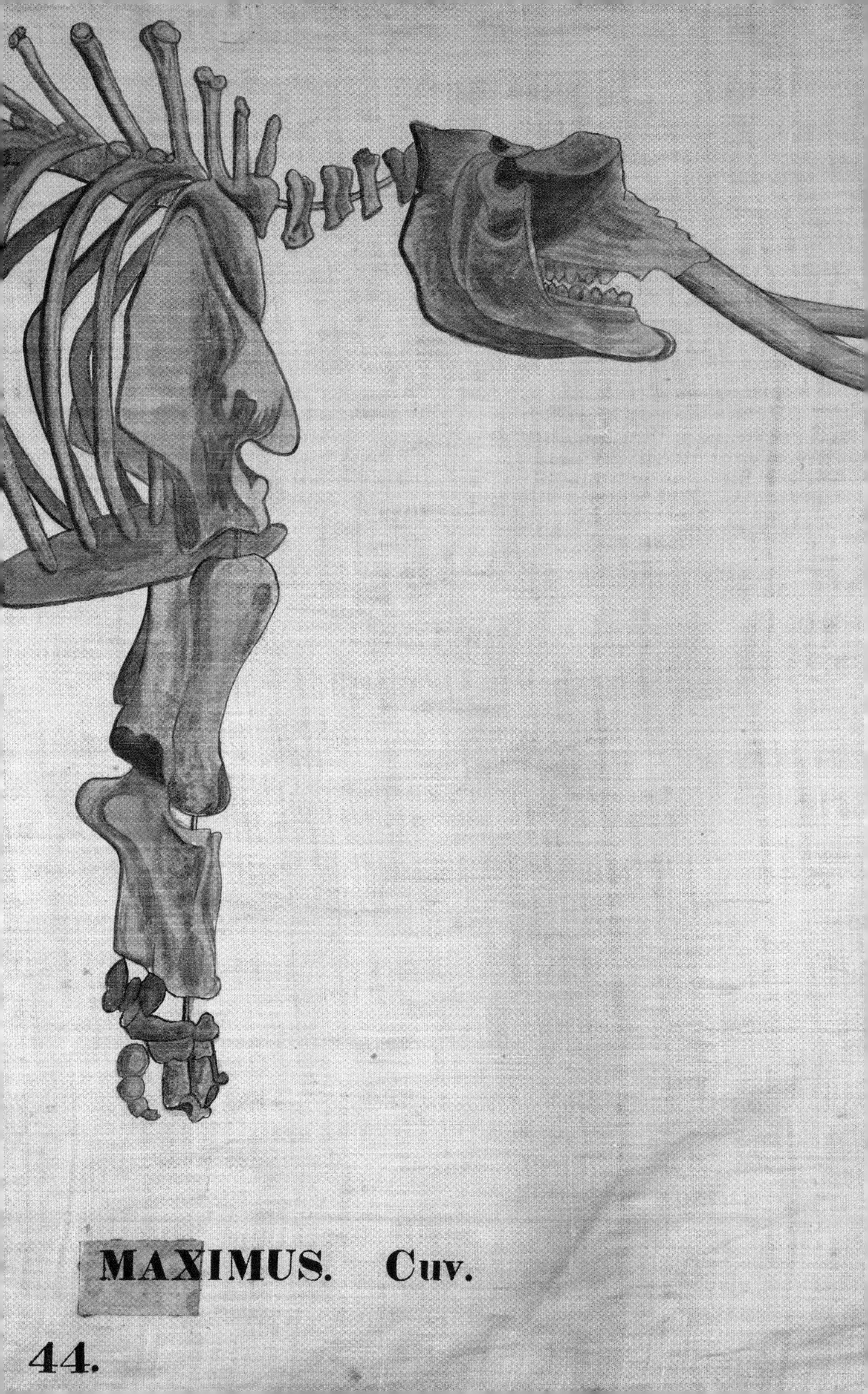
MAXIMUS. Cuv.
44.

A Family Affair

If the marriage of religion and science appeared occasionally strained in the eighteenth century, by the end of the nineteenth many believed it seemed ready for divorce. Yet the complexity of their connection endured, as perhaps can be seen most clearly in an actual married couple who together demonstrated that belief in God and trust in scientific discoveries could indeed cohabitate peacefully.

When the geologist-theologian Edward Hitchcock (1793–1864) required illustrations for reference during his lectures at Amherst College in western Massachusetts, he turned to his wife, Orra White Hitchcock (1796–1863), whose talent with pens and brushes complemented her husband's skill with exegesis and digging trowels. Her portrait of a mastodon shown here represents the pair's efforts to reconcile traditional religious understandings with the new insights into the past that science provides.

Trained initially as a Congregational minister, Edward Hitchcock left the pulpit in his early thirties to become a professor of chemistry, geology, natural history, and natural theology. Though today that may seem an unlikely portfolio, such far-ranging interests were practically required during his tenure at Amherst, from 1825 to 1854. It was a time of remarkable geological and paleontological discoveries that would dramatically revise the most basic assumptions about the age of the Earth, its composition, and its connection to widely believed biblical accounts of creation.

While teaching at Amherst, Hitchcock simultaneously served as the state geologist of Massachusetts and had leading roles in geological surveys of surrounding states. The wealth of data these efforts made available in turn influenced his theological endeavors. His 1851 book *The Religion of Geology and Its Connected Sciences* reinterpreted Christian scripture in light of new evidence his fieldwork had provided, making the then-radical claim that the seven days of creation described by the Book of Genesis may have unfolded over a period significantly longer

Drawing of a mastodon maximus skeleton (detail) by Orra White Hitchcock, ca. 1828–40.

Succeeding pages: Paleontological chart by Orra White Hitchcock, from *Elementary Geology* by Edward Hitchcock, 1840. Created nineteen years before the publication of *On the Origin of Species*, this illustration was one of the earliest depictions of a "Tree of Life" that connects all living things.

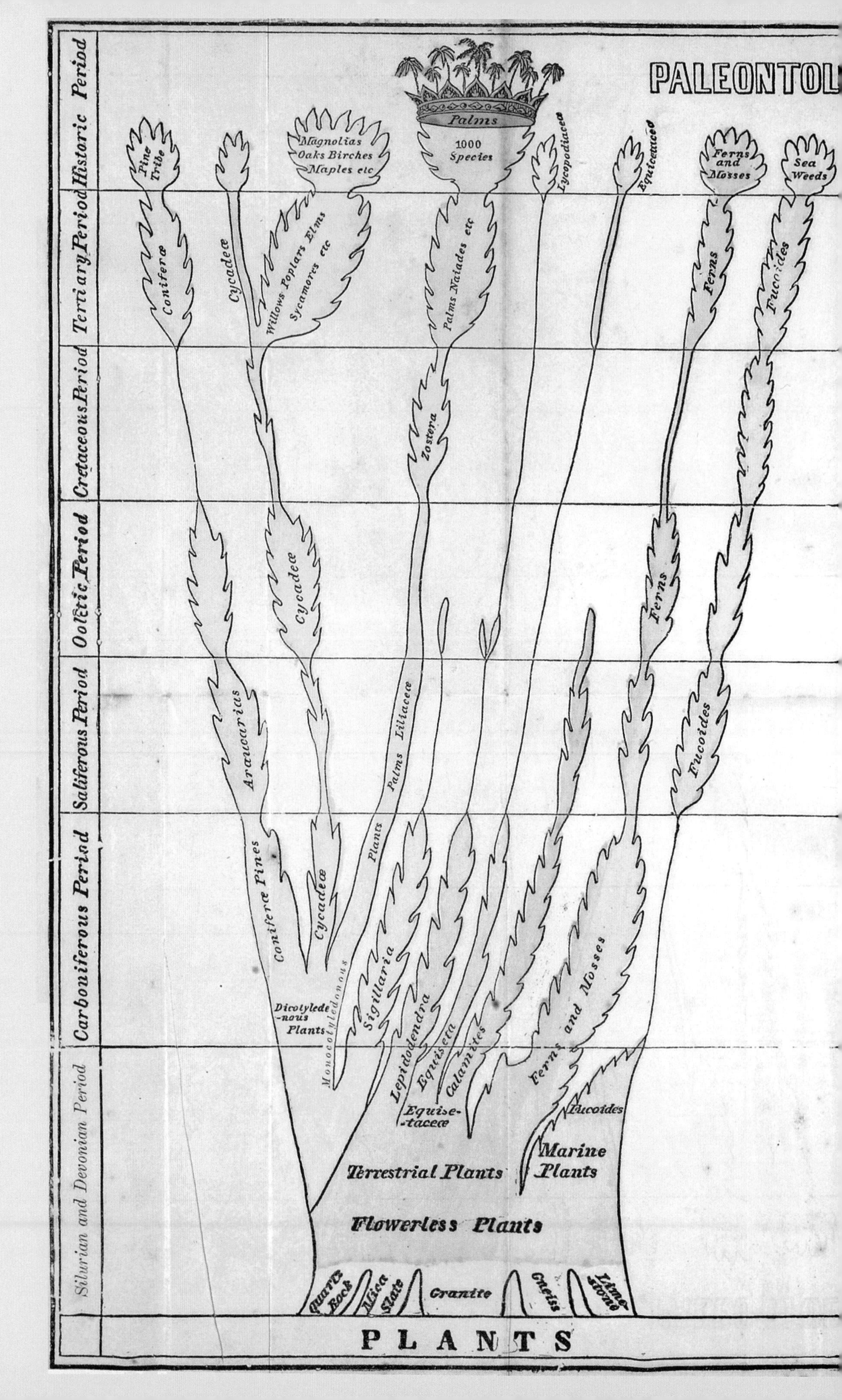
PALEONTOL
Historic Period
Tertiary Period
Cretaceous Period
Oolitic Period
Saliferous Period
Carboniferous Period
Silurian and Devonian Period
Pine Tribe
Magnolias Oaks Birches Maples etc
Palms
1000 Species
Lycopodiaceæ
Equisetaceæ
Ferns and Mosses
Sea Weeds
Coniferæ
Cycadeæ
Willows Poplars Elms Sycamores etc
Palms Naiades etc
Ferns
Fucoides
Zostera
Araucarias
Palms Liliaceæ
Plants
Coniferæ Pines
Cycadeæ
Dicotyledenous Plants
Monocotyledonous
Sigillaria
Lepidodendra
Equisetæ
Calamites
Ferns and Mosses
Equisetaceæ
Terrestrial Plants
Marine Plants
Flowerless Plants
Quartz Rock
Mica Slate
Granite
Gneiss
Limestone
PLANTS

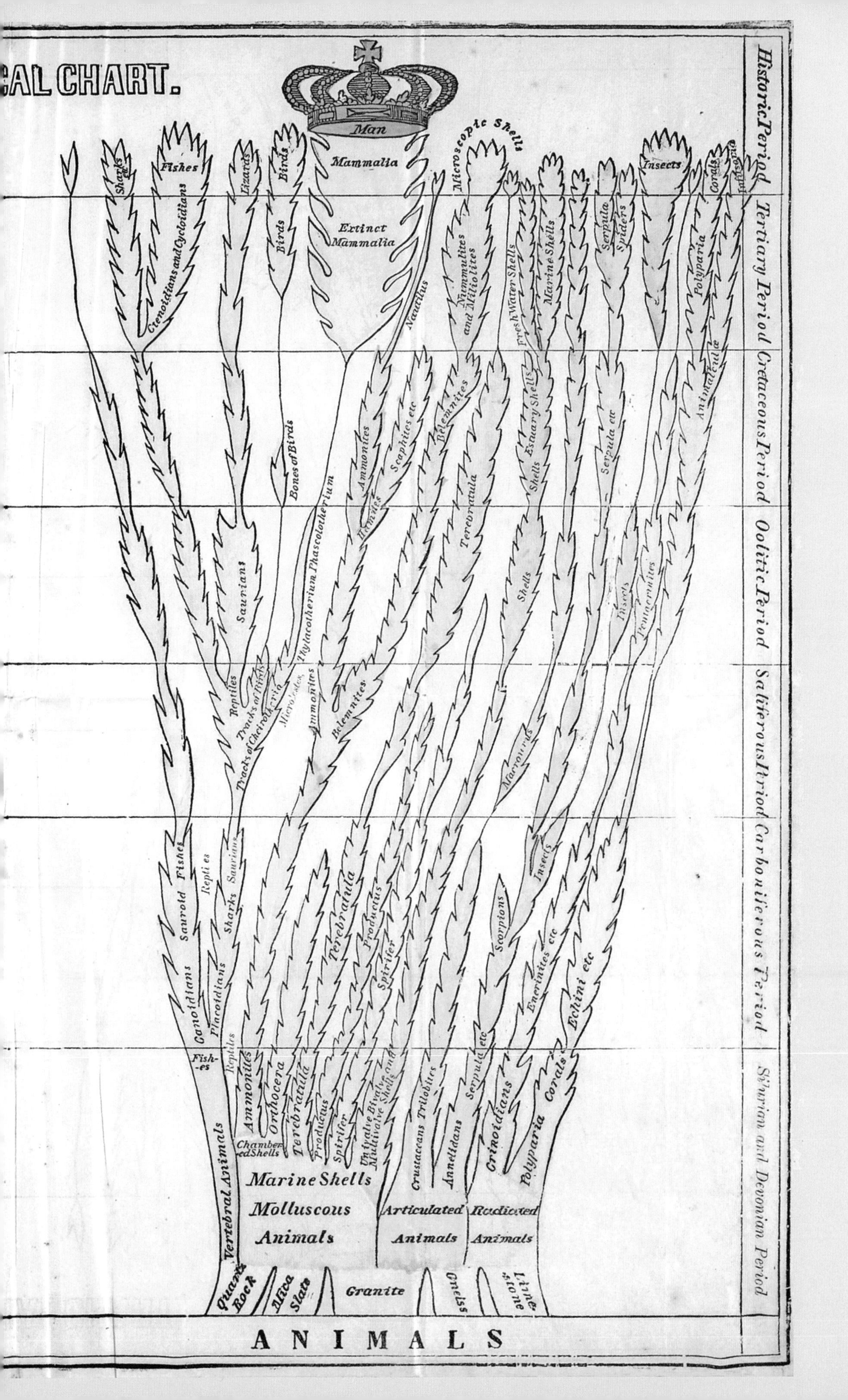

AL CHART.
Man
Mammalia
Extinct Mammalia
Microscopic Shells
Sharks etc
Fishes
Ctenoidians and Cycloidians
Lizards
Birds
Birds
Bones of Birds
Nautilus
Nummulites and Miliolites
Fresh Water Shells
Marine Shells
Serpula
Spiders
Insects
Corals
Polyparia
Animalculae
Ammonites
Scaphites etc
Belemnites
Terebratula
Estuary Shells
Shells
Serpula etc
Hamites
Saurians
Thylacotherium Phascolotherium
Shells
Insects
Pennenaites
Reptiles
Tracks of Birds
Tracks of Cheirotherium
Microlestes
Ammonites
Belemnites
Marsupials
Sauroid Fishes
Reptiles
Sharks Saurians
Ganoidians
Placoidians
Terebratula
Productus
Spirifer
Scorpions
Insects
Encrinites etc
Echini etc
Fish-es
Reptiles
Ammonites
Orthocera
Terebratula
Productus
Spirifer
Univalve Bivalve and Multivalve Shells
Crustaceans Trilobites
Annelidans
Serpula etc
Crinoidians
Corals
Polyparia
Chambered Shells
Vertebral Animals
Marine Shells
Molluscous
Animals
Articulated
Animals
Radiated
Animals
Quartz Rock
Mica Slate
Granite
Gneiss
Lime-stone
ANIMALS
Historic Period
Tertiary Period
Cretaceous Period
Oolitic Period
Saliferous Period
Carboniferous Period
Silurian and Devonian Period

than a single week. Could it be possible that each "day" was actually millions of years? Through creative readings of the original Hebrew text, Hitchcock artfully threaded the needle of not contradicting holy writ while arguing strongly for reading between the lines.

"Geology is usually regarded as having only an unfavorable bearing upon revealed religion; and writers are generally satisfied if they can reconcile apparent discrepancies," Hitchcock wrote. "But I regard this as an unfair representation; for if geology, or any other science, proves to us that we have not fairly understood the meaning of any passage of Scripture, it merely illustrates, but does not oppose, revelation."

Two hundred years after Hitchcock began his work, geology seems perhaps the least controversial of the hard sciences. When it comes to testifying to the nearly unimaginable age of the Earth, rocks seem to speak for themselves. This was far from the case in Hitchcock's day, when anyone making such claims opened themselves up for attack. "Men of respectable ability, and decided friends of revelation," he wrote, "having got fully impressed with the belief that the views of geologists are hostile to the Bible, have set themselves to an examination of their writings, not so much with a view of understanding the subject, as of finding contradictions and untenable positions." In the religious circles in which Hitchcock moved, books were written against geology, which abounded with "prejudices . . . striking misapprehensions of facts and opinions . . . severe personal insinuations" and "great ignorance." He saw his task, then, to be making his world-shaking claims as palatable as possible. One part of this strategy happened to share his home.

Orra White Hitchcock was the first woman in America to make a name for herself as a scientific illustrator. Her charts and diagrams enlivened the pages of her husband's written work, presenting the geological strata in colorful displays and giving form to theories of the relationship of various

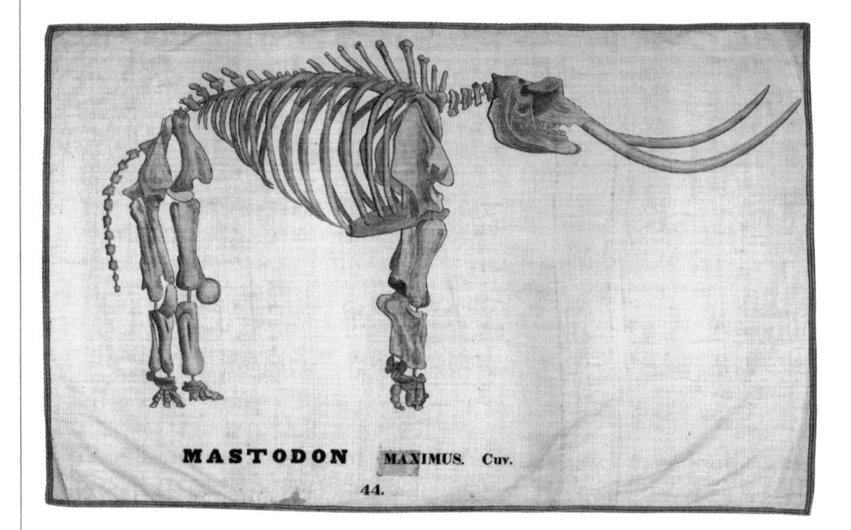

Orra White Hitchcock was one of America's first scientific illustrators. Her drawing of a mastodon skeleton was used as a teaching aid in classes on geology and natural history.

species of flora and fauna decades before Darwin introduced his theory of evolution. From the first years of Edward Hitchcock's time at Amherst, Orra White Hitchcock produced hundreds of wall-sized illustrations evocative of the banners displayed in the tents of revivalist preachers but depicting cross-sections of geological formations as well as recreations of fossils, dinosaur footprints, and prehistoric beasts. Through constant reference to these teaching tools during his lectures, Hitchcock brought to life his proposition that the discoveries of science need not call religion into question but could provide more information with which to explore its meaning.

While at Amherst, Hitchcock gathered the most significant collection of prehistoric materials in the country, including America's first confirmed fossil footprints, which had been found in Massachusetts in 1802. At the time of their discovery, locals called the bird-like tracks the footprints of Noah's raven, suggesting that one of the two birds sent from the Ark after the Genesis flood had left imprints in the mud where the waters had receded. Though only a half-serious nickname, the stewards of the collection of which the fossil became a part took seriously the need to reconcile biblical accounts of creation with geological discoveries. Eventually the latter would supplant the former as authoritative clues to gaining a better understanding of the past, but throughout the period in which Hitchcock worked, not even the most worldly and well informed would have dismissed the former so quickly.

Just as the Hitchcocks were ahead of their time in matters of science and religion, so they were in terms of their scientific, religious, and artistic partnership. *The Religion of Geology* was dedicated by its author "To My Beloved Wife," whose contributions Edward celebrated whenever he had the opportunity. The role of women in the sciences has long been overshadowed, in part because many of the so-called "great men of science" benefitted from unacknowledged contributions from their wives, sisters, and daughters. It is in this light that Edward and Orra Hitchcock's intellectual partnership stands out so notably.

"While I have described scientific facts with the pen only, how much more vividly have they been portrayed by your pencil!" Edward wrote to his wife. "And it is peculiarly appropriate that your name should be associated with mine in any literary effort where the theme is geology; since your artistic skill has done more than my voice to render that science attractive to the young men whom I have instructed. I love especially to connect your name with an effort to defend and illustrate that religion which I am sure is dearer to you than everything else."

In a time when religion and science were beginning to be seen as antagonists, Edward and Orra Hitchcock showed that—in one family at least—they could live in harmony.

Edward Hitchcock and Orra White Hitchcock, ca. 1859–60. The married couple worked together to create influential studies of geology and paleontology.

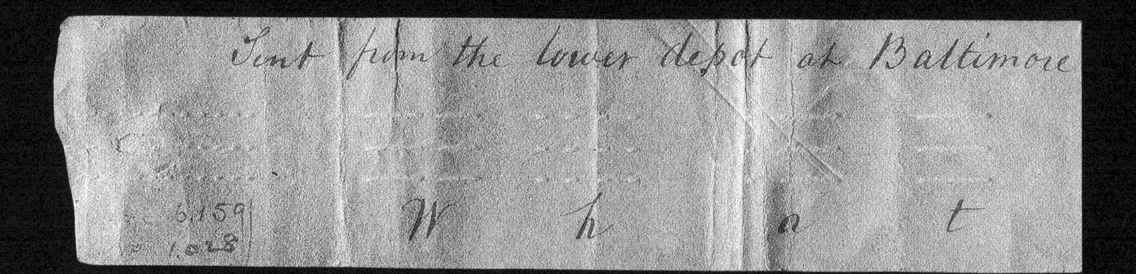

Sent from the lower depot at Baltimore

6159
1028

W h a t

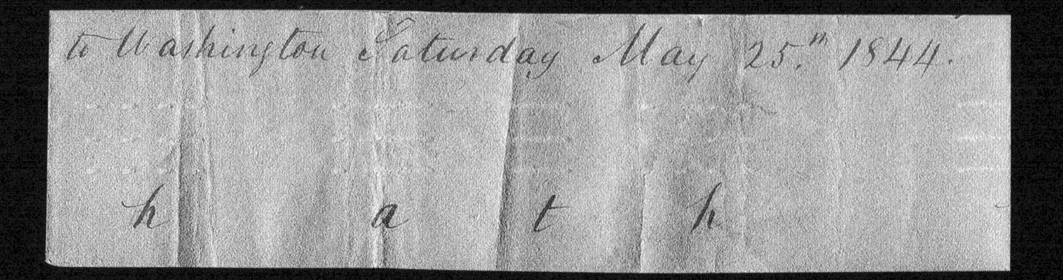

to Washington Saturday May 25th 1844.

h a t h

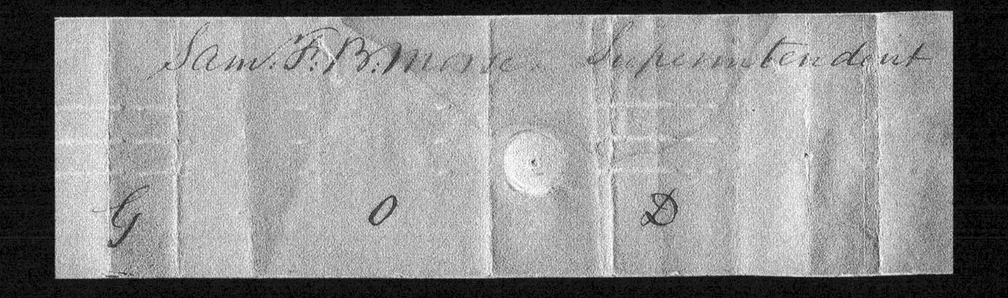

Sam.l F. B. Morse. Superintendent

G O D

of Elec. Mag. Telegraphs.

w r o u

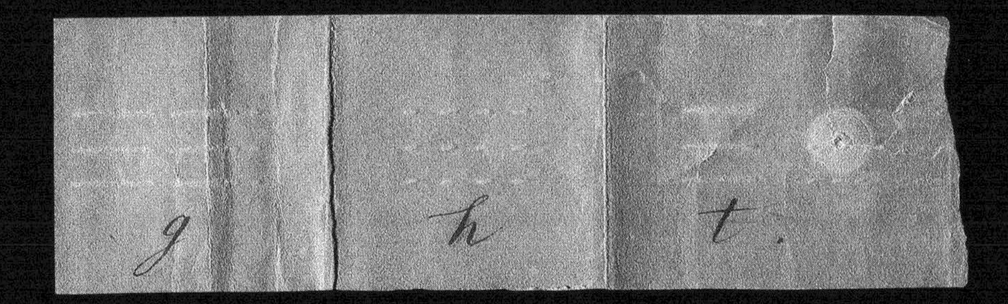

g h t.

What Hath God Wrought

On May 24, 1844, Samuel F. B. Morse (1791–1872) officially introduced the first long-distance telegraph line in the United States. After years of research and experimentation, Morse would finally demonstrate his new communication technology, which transmitted messages in the form of electrical signals over wire. Using a telegraph key, an operator could push down a small lever—briefly for a short signal, or dot, or longer for a dash—to complete a circuit formed with a receiver at the other end of the line. These electrical pulses representing letters and numbers could be sent over great distances at unprecedented speeds.

Paper telegraph tape of a message sent from Baltimore to Washington, DC, on May 24, and signed on May 25, 1844.

Morse foresaw that his invention would have a monumental impact on American society. When it came time to inaugurate his telegraph, he chose his words carefully. Sent from the US Capitol building in Washington, DC, to a railway station in Baltimore, Morse's first telegraph read, "What hath God wrought."

The phrase comes from Numbers 23:23, the fourth book of the Hebrew Bible. In this passage the Israelites have escaped oppression in Egypt and are attempting to enter the promised land of Canaan. A powerful seer named Balaam is hired to stop the Israelites with a curse, and yet to his surprise he finds that he can only utter blessings. Balaam prophesies that the Israelites will rise up like a great lion, and people will one day look at them in awe and say, "What hath God wrought!"

Morse was a committed Christian, but he did not in fact choose this message himself. The quote was suggested by Annie Ellsworth, the daughter of Morse's supporter and US Patent Commissioner Henry Leavitt Ellsworth. Ellsworth and Morse likely understood that the telegraph would revolutionize communication and culture in the United States; they less likely could have predicted that "What hath God wrought" would also christen sweeping changes to American religion.

The telegraph connected people in ways that were never before possible, creating social and intellectual networks that would foreshadow the communications revolutions of the twenty-first century. Religious organizations were some of the first to adopt the new technology. Churches across the nation celebrated the 1858 construction of the transatlantic telegraph cable connecting North America and Europe with worship services identifying the telegraph as "illustrating the providence and benevolent designs of God," as one sermon said.

"The improvements of art, and the facilities of modern traffic, are the ready avenues through which the grace of God pours the riches of mercy on a sinful world," the Rev. Joseph Copp preached. "And so in the case of this sub-Atlantic Telegraph; for whatever advantage of state, of commerce, or of letters, it may be regarded and used, God will employ it for the higher purposes of religion and humanity."

Many nineteenth-century Americans believed that the rapid technological changes of their era signaled the coming of a global religious age. The telegraph was lauded as an instrument of God that would bring humanity together, and it became a key tool for American Protestant missionaries, who understood the telegraph as a way to dazzle and persuade potential converts. The American Board of Commissioners for Foreign Missions, the largest American missionary organization of the nineteenth century, directly incorporated the telegraph into its outreach. As scholar of religion Jenna Supp-Montgomerie notes, many missionaries considered those they disparaged as having "heathen minds" to be "hard and closed to Christianity" but thought they could be moved by "the power of technology and awe-inspiring performances of it."

The telegraph had an even greater influence on religious ideas outside the mainstream. Proponents of Spiritualism—the belief that the living could connect with the dead—seized upon the ability to communicate across vast distances as proof that messages could be sent to the afterlife. Among the Spiritualist mediums who compared their abilities with Morse's work were the famous Fox sisters, whose supernatural claims had first sparked the movement. "I received letters from various places," Leah Fox claimed, "saying that [spiritual knowledge]

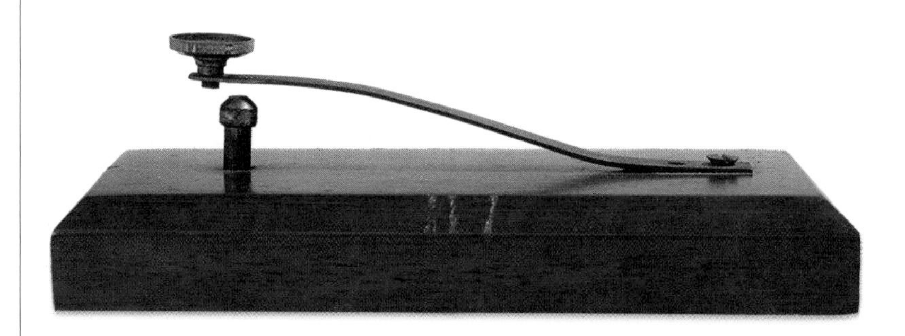

Experimental telegraph key, 1843. When the key is pushed down, it completes an electrical circuit with a receiver at the other end of the line.

had been made known through clairvoyants, speaking mediums and seers, and that the same signal had been given to all mediums. Thus we see that God's telegraph antedated that of Samuel F. B. Morse."

Samuel Morse, ca. 1845. As one of the inventors of the telegraph, Morse helped initiate a communications revolution in the United States. The telegraph would also inspire new religious movements.

Other technological advances, including photography and the harnessing of electricity, likewise excited Spiritualist imaginations. Through the phenomenon of "spirit photography," believers hoped to capture images of lost loved ones, as Mary Todd Lincoln was convinced she had in 1872, when a photographer presented her with a spectral image of her late husband.

Scientists were not unknown among the ranks of Spiritualists. In the 1850s a renowned chemistry professor named Robert Hare invented a device he called a "spiritoscope" to detect fraud among the many professional mediums of the day. His efforts to debunk the practice unexpectedly led to his own conversion to beliefs he had sought to discredit.

For the prominent Spiritualist figure Andrew Jackson Davis, telegraphy was a revolutionary way to communicate not just with the living but also the dead. Davis organized séances where individuals would arrange themselves in a circle and hold a magnetic cord. Such a cord, Davis and other Spiritualists believed, would help to transmit vital electrical signals between the realms of the living and the dead. For Spiritualists, telegraphy was not merely an earthly technology; it was a method for tapping into a sacred network of bodies and spirits that spanned the chasm separating the mundane physical world from boundless eternity. It is no surprise that the movement's first periodical was titled *The Spiritual Telegraph*.

Spiritualists were not the only adherents to incorporate the new technology of the telegraph into religious practice. The utopian Oneida Community of upstate New York, for example, celebrated the construction of the transatlantic telegraph cable, seeing it as a vehicle of instantaneous global communication that would inevitably spread their way of life out to the entire world.

But none provided a better example of the ways the telegraph bridged religion and science more than Morse himself. Late in life he used his fame and fortune to promote the idea that scientific and religious developments occurred in relation to each other. In a gift to Union Theological Seminary in New York, he endowed a lecture series on "the relation of the facts and truths contained in the Word of God, to the principles, methods and aims of any of the Sciences," a program that continued for fifty years after his death.

SOLAR
SYSTEM
E.H. BAKER
A.D. 1876

Our Celestial Visitant

Comets have always been a source of wonder. Well into the nineteenth century, they were also a source of religious questioning for many Americans. When a comet's appearance in 1843 coincided with the date self-proclaimed prophet William Miller (1782–1849) had chosen for the end of the world, Miller's followers (known as Millerites) caused a panic across the nation. Unexpectedly, this period of spiritual excitement brought about renewed interest in the science of comets, leading lecturers to include slides of the great comets of history in their magic-lantern shows.

Where some saw hysteria in the comet craze, others found opportunity. Fascination with the "Great March Comet of 1843" led to a surprise boom time in public support for science. Cities and towns of all sizes hosted experts on the subject in public meeting houses, including churches, while in Boston a fundraising campaign fueled by the desire to offer scientific answers to metaphysical queries helped improve one of most significant observatories in America (at Harvard College) by purchasing a telescope of the best quality then available.

The indirect author of much of this nascent astronomical interest, William Miller was a New York farmer and a Baptist lay preacher during the religious revival known as the Second Great Awakening. Miller had studied the prophecies of the biblical prophet Daniel and, using a specific method of symbolic analysis, predicted that the Second Coming of Jesus Christ would occur on March 21, 1843. Just weeks before the anticipated date, the comet appeared in the sky and eventually grew so bright that it could be observed in broad daylight. While this bolstered Miller's prediction, the comet eventually passed. The Millerites recalculated that Jesus would actually return seven months later, on October 22, 1844. When this day, too, came and went, the movement experienced an event that came to be called the Great Disappointment, after which many

"Solar System" quilt by Ellen Harding Baker, 1876.

Satirical cartoon depicting the ascension to heaven of the followers of self-styled prophet William Miller, 1844. Miller predicted that the world would end on October 22, 1844. Weeks earlier, a comet appeared in the sky, which helped to bolster Miller's prediction.

followers left Millerism and Christianity altogether.

Fabricated far from the scenes of the Millerites' failed rapture, a more hopeful comet can be seen on the "Solar System" quilt made by Ellen Harding Baker (1847–1886). Baker was an educator and astronomer whose lectures provided science education to public audiences throughout Iowa in the 1880s. Answering the local desire for lectures on the movements of the planets and the position of stars in the night sky, Baker put her talent for quilting to work, producing large-scale visual elements to accompany astronomy lectures held for the public.

Constructed of wool, cotton, and silk embroidery, the appliqué quilt was begun in 1876, as seen in the date indicated along with the creator's name in the lower corners. In 1883 newspaper reports around the country announced, "Mrs. M. Baker, of Lone Tree, Iowa, has just finished a silk quilt upon which she has been at work seven years, in which is worked the solar system. She went to Chicago to view the comet and sun spots through the telescope that she might locate them accurately." Baker highlighted the sun at the center of the quilt and stars at its outer edges. The five small dots depicted closest to the sun are Mercury, Venus, the Earth and its moon, and Mars. Beyond these first concentric rings are the larger planets: Jupiter, Saturn, Uranus, and Neptune, each with the respective numbers of moons or rings known at the time.

Perhaps the quilt's most eye-catching detail is the comet that can be seen streaking into view from the top left corner. Historians argue that Baker was likely depicting one of three comets that had made a splash in nineteenth century America: There was the Great Comet of 1843 that had for the Millerites heralded the end of the world. In 1874 another comet known as Coggia appeared in the night sky and stayed there for weeks; its long, flickering tail undulated so rapidly that one observer likened it to a "fine gauze wavering in a strong breeze." The final possibility is the Great Comet of 1882, which caused a stir around the country and would have appeared in the Iowa sky during the final months of the "Solar System" quilt's completion.

As a publication of the Iowa State Historical Society expressed it several decades later, Iowans had a natural affinity for comets, born perhaps of their relationship to the open sky. "Like the shepherds of yore, the early pioneer had time to know and appreciate the wonders of the heavens," the historian Ben Hur Wilson wrote in 1940. "To him they had a practical value, for by the stars at night he often found his way across the trackless prairies, and by the sun he told the hours of the day. Whether the motive was religious inspiration, belief in the soothsayings of astrologers, or devotion to the science of astronomy, people stood in wonderment as they beheld the glories of the midnight sky. Of the many awe-inspiring spectacles in the heavens, none was more likely to arouse sinister forebodings among people than the appearance of a comet."

Throughout the months of the quilt's contruction, Iowa newspapers offered detailed descriptions of the comet's approach. "The antics of the comet are exciting the astronomers," the *Oskaloosa Herald* reported on October 19, 1882. The previous month, the paper noted, telescopes had provided an indication that "a separation was seen of the nucleus," the ball from which the comet's tail extended. The play-by-play commentary that followed would have well served a sporting event. Not to be outdone, the *Iowa City Republican* declared that the "present comet in the Eastern sky, which can be distinctly seen by everyone at early morning, is certainly the most remarkable of all the modern comets."

No press report rivaled the florid coverage found in the *Iowa State Register*, whose resident poet penned "To Our Celestial Visitant":

Mysterious wanderer of the sky In heaven's bright caravan, In thy celestial journey, why Seek'st thou the abode of man?

After twenty lines, the verse concluded:

Flashing through the realms of space, What other beings gaze on thee, And in thy wondrous grandeur trace The matchless hand of Deity?

It is not known if Baker happened upon these lines as she completed her seven years labor on the "Solar System" quilt, but she, too, sought to give expression to the meeting point of spiritual wonder and scientific observation, allowing all who saw her quilt to gaze upon a vision of otherworldly awe caught in a humble presentation of cotton and wool.

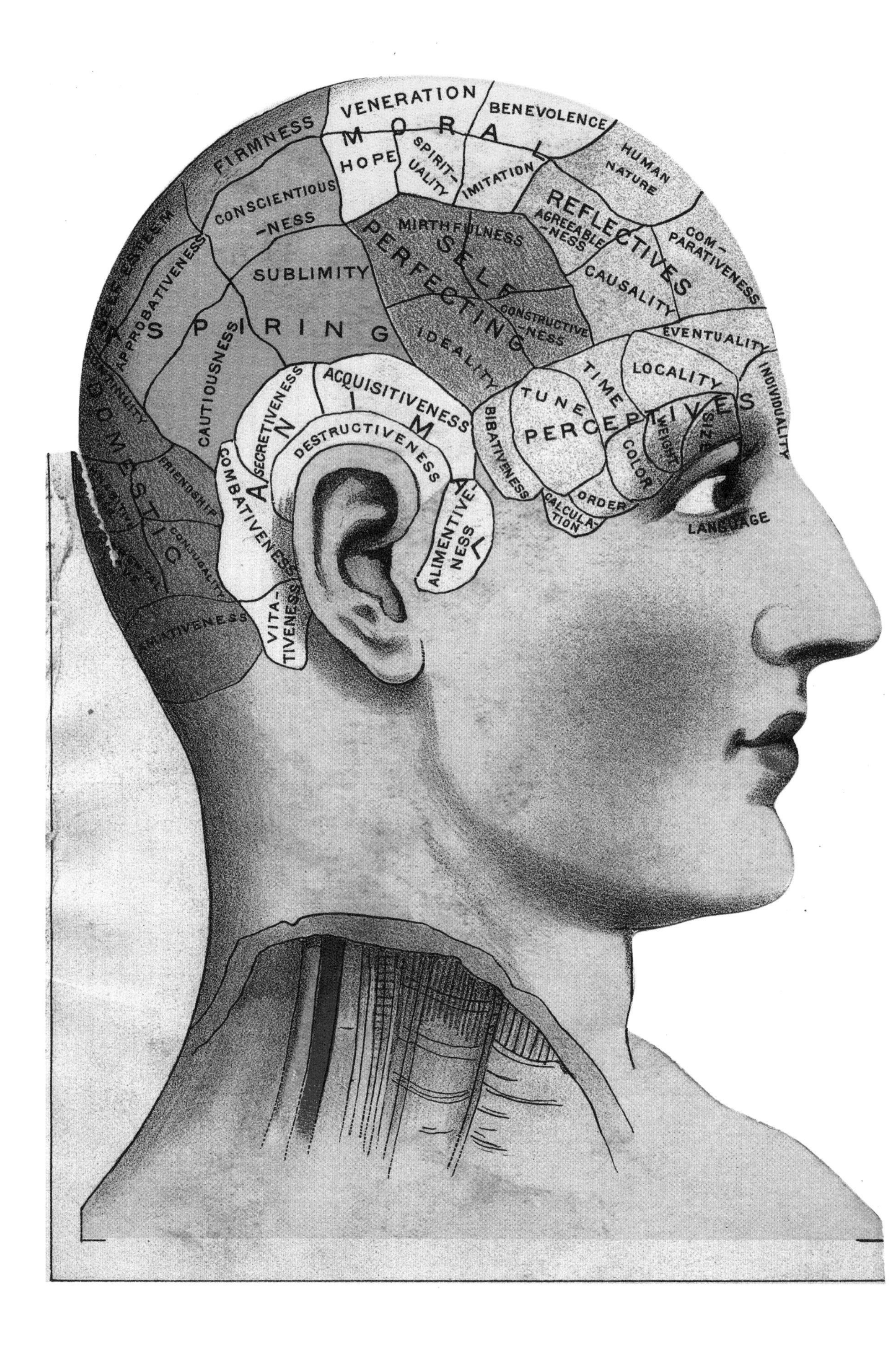

VENERATION
BENEVOLENCE
MORAL
FIRMNESS
HOPE
SPIRIT-UALITY
IMITATION
HUMAN NATURE
CONSCIENTIOUS-NESS
REFLECTIVES
AGREEABLE-NESS
COM-PARATIVENESS
MIRTHFULNESS
SELF PERFECTING
SUBLIMITY
CAUSALITY
APPROBATIVENESS
ASPIRING
CONSTRUCTIVE-NESS
IDEALITY
EVENTUALITY
CAUTIOUSNESS
TIME
LOCALITY
ACQUISITIVENESS
ANIMAL
TUNE
PERCEPTIVES
INDIVIDUALITY
SECRETIVENESS
DESTRUCTIVENESS
BIBATIVENESS
WEIGHT
SIZE
COLOR
DOMESTIC
FRIENDSHIP
COMBATIVENESS
ORDER
CALCULA-TION
LANGUAGE
ALIMENTIVE-NESS
CONJUGALITY
VITA-TIVENESS

The Anatomy of the Soul

What if you could hold a person's soul in your hands and measure it? What if, tracing your fingers across its delicate surface, you could understand everything about that person's personality? This was the dream of phrenology, a now-debunked brain science that originated in Europe and then took America by storm during the nineteenth century.

Phrenology is best remembered for its maps of the mind. In pamphlets, illustrated busts, and vivid anatomical maps, Americans encountered something akin to an atlas of the soul. Phrenological maps depicted areas not only for basic cognitive functions, like language or the perception of time, but also more abstract qualities like "conscientiousness," "veneration," and "spirituality."

Phrenology was initially developed by Viennese physician Franz Joseph Gall (1758–1828) in the late 1700s. Gall was a pioneer in what would later be called neuroanatomy, known for his innovative methods for dissecting the human brain. Where his contemporaries sliced across the top of the brain and observed the cross section, Gall would turn the brain upside down in his hands and methodically splay it open like a butcher. Working from the brain stem upward, Gall would trace the serpentine paths of nerve fibers as they exited the spinal cord and tunneled into dense little "bundles of threads" that he found packed throughout the skull.

Gall hypothesized that these bundles of thread functioned like mental organs, each responsible for a different mental function. He eventually mapped out twenty-seven distinct regions of the human brain, ranging from "mechanical skill" to "vanity" to "theosophy," which Gall defined as a "sense of God and Religion." Gall maintained that the size of these mental organs corresponded to the development of that trait within a person's personality. Thus a person with a large "secretiveness" area, for example, would be more likely to conceal

Illustration mapping phrenological regions of the skull from *The Household Physician* by Hebert E. Buffum et al., 1905.

their intentions; a small "theosophy" area might explain why a person struggled with their faith.

If phrenology had been limited to these tenets it likely would have been forgotten. But the Austrian physician made one additional assumption that would help turn phrenology into a heresy at home and a social phenomenon in the United States. Over a series of public lectures during the late 1790s, Gall argued that the skull takes the shape of the brain beneath, and that a person's entire personality—their behaviors, beliefs, aptitudes, and even their moral character—could therefore be measured according to the bumps and ridges of their head.

This was a revolutionary idea. The human mind had classically been studied only through introspection. Gall was trying to replace cogitation with the caliper. If successful, it would mark a significant scientific accomplishment, transforming the subjective realm of the mind into something objectively knowable. But could consciousness—that divine spark believed to separate us from animals—really be made of mere flesh? Could our personalities, beliefs, and morals all be reducible to brute matter?

Phrenology sparked immediate controversy. Many were skeptical of Gall's scientific reasoning; even more were offended by phrenology's materialism. To reduce religiosity to anatomy seemed inherently atheistic. If the pious worshipper is motivated by an enlarged brain region, what room would there be for God? Phrenology was rejected by the Roman Catholic Church, and Gall was banned from publishing or lecturing on the topic ever again. The ban, signed by Holy Roman Emperor Francis II in December 1801, argued that phrenology was materialistic and therefore violated "the first principles of morality and religion." The decree listed additional reasons for prohibiting Gall's public lectures, including the fact that women had been witnessed in attendance.

Being labeled a heretic earned Gall an initial wave of publicity abroad, and he was invited to speak in universities and courts throughout Europe. He left Austria in 1805 on a public lecture tour that would never end, racing across more than fifty cities amidst the chaotic violence of the Napoleonic Wars. Nevertheless, when Gall died in Paris in 1828, his neuroanatomical ideas had not gained widespread purchase. In most countries phrenology was immediately rejected; in England it garnered initial enthusiasm before being abandoned due to its fundamental problems as a scientific theory.

In the United States, however, phrenology would soon become a cultural phenomenon. In 1832 Gall's protégé Johann Spurzheim (1776–1832) visited America on a whirlwind lecture tour that converted thousands. Spurzheim was heralded by many prominent thinkers as a genius; Ralph Waldo Emerson said he was one of the world's greatest minds.

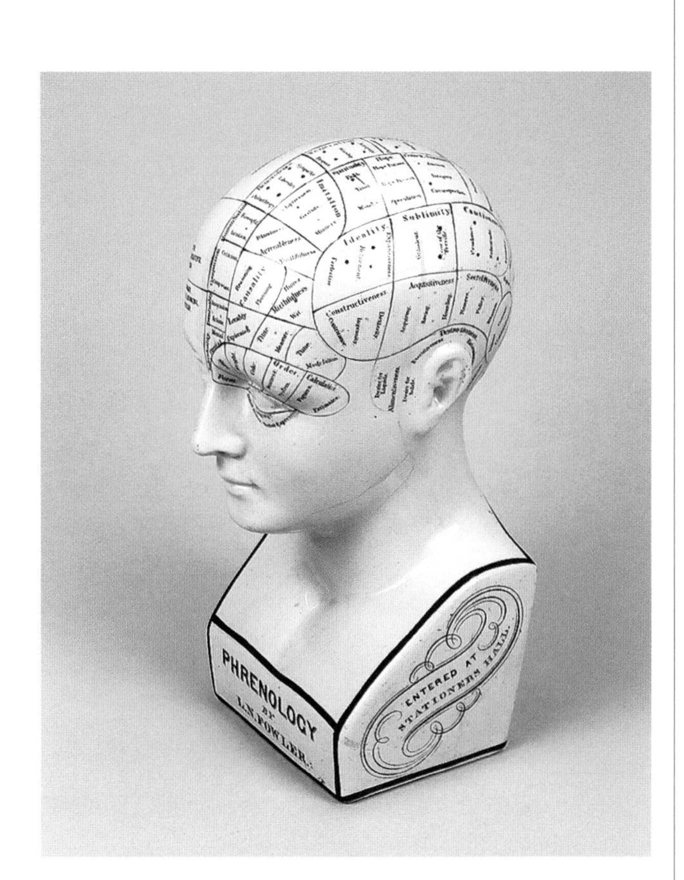

Phrenology bust by Lorenzo Niles Fowler, ca. 1855. Phrenologists believed that bumps on the skull corresponded to personality traits, such as "agreeableness" and "spirituality."

After only six months in the United States, Spurzheim contracted typhoid fever and died in Boston. Thousands attended the funeral and, after a public autopsy, Spurzheim's brain was preserved as a relic and put on public display.

Phrenology quickly became a nationwide craze. Everyone from Walt Whitman to Brigham Young to P. T. Barnum sat for readings, and eminent literary figures like Herman Melville and Edgar Allan Poe incorporated phrenology into their works. It was embraced by doctors, scientists, and religious leaders alike and briefly swept through the leadership of the Unitarian Church. Employers asked for phrenological profiles when interviewing workers. Women developed new hairstyles to show off their best phrenological features.

Why did phrenology take off in the United States after fizzling in most European countries? One reason was that by the mid-nineteenth century it seemed less controversial to posit anatomical explanations of behavior. The nineteenth century was a period of revolutionary new understandings of life. With the discovery of the cell, the rise of organic chemistry, and advances in physiology and biomechanics, it was becoming increasingly clear that living things were made up of smaller physical mechanisms. If plants and animals are composed of smaller parts that function together to create a single entity, perhaps the same could be said of the human mind. The notorious case of Phineas Gage, an American railroad foreman whose personality was changed after an iron rod was driven through his skull, lent further credence to the notion that the brain was made of smaller units.

Phrenology's success in the United States was also due to the work of three enterprising siblings: Orson Squire Fowler, Lorenzo Niles Fowler, and Charlotte Fowler Wells. Orson, the oldest of the three, was introduced to phrenology through Spurzheim and considered it a revelation. He convinced his siblings to leave their jobs and join him in his calling, and together they created a craniological empire that would last for more than three decades. The Fowler brothers toured the country to spread the "truth of phrenology," offering readings for two cents. They published dozens of books and measured the skulls of everyone from Henry Ward Beecher to then-congressman James A. Garfield, who kept a dictation of Lorenzo Fowler's reading in his diary. (Garfield's brain, Fowler diagnosed, was "very large—too large," with a "remarkable power of accumulating knowledge.")

Thousands of Americans sat for phrenological readings, closing their eyes to feel an expert's fingers trace the truths written across their skulls.

Charlotte Fowler Wells, meanwhile, helped transform the family venture into an institution. She founded and managed their phrenological publishing house, Fowler & Wells Company, where she also served as a writer, editor, proofreader, and educator. Wells oversaw the company's Phrenological Cabinet, a public museum in New York that displayed phrenological portraits of hundreds of famous Americans. The Phrenological Cabinet became a popular local attraction where visitors could discover the truth behind America's most eminent and infamous minds. After the death of Aaron Burr in 1836, the Fowlers ordered a cast of his head that was quickly put on display. The bumps on Burr's head supposedly revealed an unusually large region of "Destructiveness."

The Fowlers were not just savvy promoters of phrenology; they also adapted the controversial new science for American sensibilities. Orson and Lorenzo had both studied for the ministry, and they argued extensively that phrenology could in fact reconcile religion and science: "If revelation and phrenology are both true, there must be a perfect harmony and coincidence between the theology of phrenology, and the theology of revelation.... Both together, would give a far more perfect view of theology and religion, than either can do separately."

Rather than explaining religion away, in other words, phrenology could go hand in hand with theology. In making this argument, the Fowlers joined a long tradition of Americans—including Benjamin Franklin, Cotton Mather, and Joseph Priestley—who believed that scripture could be complemented by the so-called Book of Nature. If nature's immutable laws were authored by God, then a science of the mind would only further illuminate God's words and works.

The Fowlers also cast phrenology as the ultimate tool for self-improvement, capitalizing on America's unique fascination with personal development. Phrenology, they argued, allows the individual to measure their cognitive weaknesses and thereby improve upon them. Foreshadowing "brain training" and the concept of neuroplasticity, the Fowlers emphasized that the brain could be changed. Even religion, they argued, could be cultivated using simple techniques: "Study and admire the divine in nature . . . ; cultivate admiration and adoration of the Divine character and government, of this stupendous order of things, of the beauties and perfections of nature, as well as a regard for religion and things sacred."

If it were used to better contemplate the divine, what could be wrong with a braille of the soul? The Fowlers helped turn phrenology into an American phenomenon by promising that, with just a pamphlet and two pennies, anyone could receive scientific insights into their character. The Fowlers transformed phrenological reading into a spiritual practice. Thousands of Americans sat for such readings, closing their eyes for a few minutes to feel an expert's fingers trace the truths written across their skulls.

As privileged white Americans were measuring their minds for the purposes of self-improvement, other groups of people were being measured for brutal purposes. Scientists in the United States and abroad used phrenology to justify the enslavement of Africans and the conquest of Native American lands. Kentucky physician Charles Caldwell (1772–1853) claimed that the skulls of Africans had larger "Veneration" and "Cautiousness" regions, indicating that they were mentally suited to servitude. Philadelphia physician and scientist Samuel George Morton (1799–1851) used phrenology to support polygenism, the racist and false theory that the human races were separate species.

Phrenology left behind a complicated legacy in the United States. It was a pseudoscience, yet it was also a pioneering approach to the study of the mind that predated key elements of modern cognitive psychology. It was radically materialistic, and yet it also inspired new religious ideas and became a source of meaning for an entire generation. It was used by some to claim the superiority of the European race, and yet it was also widely cited in arguments for criminal reform and the abolition of slavery. It sometimes masqueraded as an anatomical justification for misogyny, and yet its emphasis on embodiment helped question gendered assumptions about the human body. Phrenology is rightly remembered as a failed theory, and yet it was also something more: a cultural phenomenon that blurred the boundaries between the mind and brain, the body and the self, and religion and science.

Darwin in America

Though his theory of evolution by natural selection was controversial during his lifetime, no one could have predicted the enduring cultural significance of Charles Darwin (1809–1882) himself or the starring role he would play in the perceived conflict between religion and science more than a century after his death. Yet the most intriguing aspect of Darwin's influence on American culture may be that, initially, his groundbreaking insights were embraced by many who also wholeheartedly endorsed the Bible.

Darwin's landmark 1859 book, *On the Origin of Species by Means of Natural Selection, or the Preservation of Favoured Races in the Struggle for Life*, is considered the foundation of evolutionary biology. Using meticulously collected evidence, Darwin theorized that the diversity of life-forms on the planet was due to a process of natural selection. In the struggle for survival, Darwin held, certain small, heritable variations in a population can make an individual more suited to its environment and therefore more likely to reproduce. This process "selects" some traits over others and ultimately causes populations to change over time.

On the Origin of Species proposed a radically new understanding of the forces responsible for the variety of life-forms on Earth. Darwin described his theory using the metaphor of a "tree of life":

The green and budding twigs may represent existing species; and those produced during each former year may represent the long succession of extinct species. . . . As buds give rise by growth to fresh buds, and these, if vigorous, branch out and overtop on all sides many a feebler branch, so by generation I believe it has been with the great Tree of Life, which fills with its dead and broken branches the crust of the earth, and covers the surface with its ever branching and beautiful ramifications.

Portrait of Charles Darwin (detail) by Julie Margaret Cameron, ca. 1870. After his theory of natural selection revolutionized the life sciences, Darwin himself became an emblematic figure in the perceived conflict between science and religion.

Title page from Charles Darwin's *On the Origin of Species*, 1859. Darwin's theory of evolution became a lightning rod issue in debates about religion and science in American public life.

Darwin's theory of natural selection was controversial, in part, because it offered a naturalistic explanation for the varieties of animal life. Where natural theologians argued that the existence of so many species—each somehow perfectly suited to their environments—was proof of divine planning, *On the Origin of Species* offered an earthly, almost mechanical explanation. God, it seemed, was no longer necessary to explain creation. Many of Darwin's supporters claimed that evolutionary theory would be a death blow to religion. Critics argued that Darwin's ideas were inherently atheistic.

Yet, after the publication of *On the Origin of Species* in 1859, Darwin's most ardent supporter in the United States turned out to be Asa Gray (1810–1888), a devout Presbyterian who was also one of the most influential American scientists of the nineteenth century. As a professor of botany at Harvard University, Gray had extended and unified the taxonomy of North American plants. He would eventually become one of Darwin's closest intellectual partners and friends. Although Gray and Darwin disagreed on key theological points, their continued relationship is an example of the complex relationship between religion and science in America.

Darwin first wrote to Gray asking for information on American flowers that would ultimately help inform his theory. For his own part, Gray had already independently begun to question prevailing biological assumptions of his day. He avoided talk about abstract biological types and had come to see plant species as fluid units that could not be defined within strict boundaries. Historians contend that Darwin's thinking about evolution was influenced by Gray's studies of the varieties and distribution of plant species.

Gray helped to arrange the publication of the first American edition of *On the Origin of Species* in 1860. He quickly became the book's most famous and widely read commentator in the United States, and he was particularly forceful in his support of Darwin against Louis Agassiz (1807-1873), the preeminent scientist of the era, whose arguments against Darwinian evolution included racist insistence that Africans and Europeans had been separately created by God. In Agassiz's estimation, the Book of Genesis presented a true account of human origins, but those origins only included the lighter shades humanity; other races were considered by him to be different species, or as he put it, each was a different "thought of God."

Countering polygenism as poor science and faulty theology, Gray wrote a number of influential tracts on the possibility of reconciling evolutionary theory with theistic belief, aiding the reception of Darwinism in America. A common theological objection to Darwinism was related to the problem of evil: how could an omnipotent and

ON

THE ORIGIN OF SPECIES

BY MEANS OF NATURAL SELECTION,

OR THE

PRESERVATION OF FAVOURED RACES IN THE STRUGGLE FOR LIFE.

BY CHARLES DARWIN, M.A.,

FELLOW OF THE ROYAL, GEOLOGICAL, LINNÆAN, ETC., SOCIETIES;
AUTHOR OF 'JOURNAL OF RESEARCHES DURING H. M. S. BEAGLE'S VOYAGE ROUND THE WORLD.'

LONDON:
JOHN MURRAY, ALBEMARLE STREET.
1859.

Darwin's theory of evolution inspired a variety of religious responses. Asa Gray (left) was a devout Christian, a leading scientist, and an influential advocate for Darwin. Rabbi Kaufmann Kohler (center), a prominent religious voice in favor of evolution, argued that the theory was in keeping with Jewish tradition. Geologist Louis Aggasiz (right) rejected evolution in favor of creationism and the racist theory of polygenism.

benevolent God allow life to unfold through carnage and extinction? Gray turned this argument on its head:

> **Darwinian teleology has the special advantage of accounting for the imperfections and failures as well as for successes. . . . It explains the seeming waste as being part and parcel of a great economical process.**

For Gray, Darwinism helped to explain the existence of evil, in other words, by giving it a higher purpose.

Gray also believed that evolutionary adaptations reflected the intent of a Creator. Favorable variations were generated within organisms by God, he argued, and those variations were spread through the process of natural selection. While other creative evolutionists sought to make God the selecting agent, for Gray it was the objective, mechanical process of natural selection that winnowed down variations and thereby tailored organisms to their surroundings.

It was not just Christians who were inspired by Darwin to ask new questions of traditional views. After the publication of the follow-up to *On the Origin of the Species*, 1871's *The Descent of Man*, numerous religious groups published hostile responses. Many American Jewish rabbis and intellectuals were initially opposed to Darwin's depiction of human origins, focusing on humanity's unique morality and intelligence. This changed in 1874 when the German-born Reform Rabbi Kaufmann Kohler (1843–1926) gave a series of sermons on science and religion. He argued that the importance of Genesis was in its spirit rather than its specific claims about the origins of humankind. Kohler also posited that evolutionary theory was well established by natural evidence and that it supported Reform Judaism's doctrine of progressive revelation. Within a few months, several Reform rabbis who had previously opposed evolution changed their views and supported Kohler's ideas, providing another example of how religious communities in America have actively engaged with evolutionary theory over time.

Two years after Kohler began his influential lectures, Gray published *Darwiniana*, a collection of writings that argued for a conciliation of Darwinian evolution with the tenets of theism. Gray argued that science and theology could go hand in hand. Like many American scientists before him, Gray considered scientific inquiry to be in line with religious practice. For him, unveiling the laws and processes of nature was an extension of humanity's "divine desire to know." Religious traditions also answered this desire, but, importantly for Gray, religion and science did not aspire to provide the same kinds of information. "We may take it to be the accepted idea that the Mosaic books were not handed down to us for our instruction in scientific knowledge," he wrote, "and that it is our duty to ground our scientific beliefs upon observation and inference, unmixed with considerations of a different order."

Gray and Darwin did not agree on everything. The two discussed faith and religion in a series of letters from 1860 to 1862. Gray argued that adaptations required God's loving intervention. Darwin, on the other hand, wrote to Gray, "I cannot honestly go as far as you do about design." Darwin did confide that he was still perplexed: "I cannot think that the world, as we see it, is the result of chance; & yet I cannot look at each separate thing as the result of design." Darwin and Gray nonetheless remained friends and intellectual partners. Darwin considered Gray to be his best advocate, and Gray wrote that Darwin's death was "like the annihilation of a good bit of what is left of my life."

Evolution is often thought to be opposed to monotheistic religion, but the scientific theory was met in America with simultaneous theological objections and support.

In popular coverage of controversies surrounding science and religion, evolution is often thought to be opposed to monotheistic religion, but Gray and Kohler exemplify how the scientific theory was met in America with simultaneous theological objections and support. Gray was one of many scientists who considered scientific inquiry to be in line with religious belief.

"We cling to a long-accepted theory, just as we cling to an old suit of clothes," Gray said in defense of Darwin's theory. "Granting the origin to be supernatural, or miraculous even, will not arrest the inquiry."

Quilting the Cosmos

Truths can be printed in ink, but they can also be recited, danced, or woven into fabric. One of America's most innovative quilt makers was Harriet Powers (1837–1910), an African American folk artist and farmer who was born into slavery in rural Georgia. Powers's 1886 Bible quilt is considered a masterpiece of nineteenth-century quilt making. Composed of 299 separate lengths of fabric arranged into 11 motifs, the Bible quilt is a stunning patchwork of Bible stories, slave spirituals, and astronomical events that Powers recorded during her lifetime.

Powers was known to explain the Bible quilt to anyone who was interested, detailing her techniques, fabric choices, and the complex symbolism of her work. You might take as guidance the following excerpts from panel descriptions by quilt historian Kyra E. Hicks. They are based on notes taken by Powers's contemporaries during these technical conversations, so at least some of Powers's voice might be discerned: "Adam and Eve naming the animals, including a camel, elephant, leviathan, and ostrich, in the Garden of Eden," as shown on page 93. "Adam and Eve are also being tempted by a crawling serpent," and "Judas Iscariot surrounded by thirty pieces of silver, his payment for betraying Jesus. The pieces of silver were originally made of green calico fabric. The largest circular object is a star 'that appeared in 1886 for the first time in three hundred years,'" as shown on page 92.

Quilting is an ancient technology, dating at least as far back as 100 BCE. At its simplest, it is a method of stitching together layers of fabric and padding to create large, warm bed coverings using smaller pieces of assorted cloth. Quilting has existed in many cultures, often within a domain of practical domestic work cordoned off for women. Yet quilting is rarely performed out of necessity; there are easier ways to make blankets. Quilting is precise, resource-intensive work that requires expertise, and, like other crafts, it is also almost inherently expressive of

Bible quilt by Harriet Powers, 1885–86.

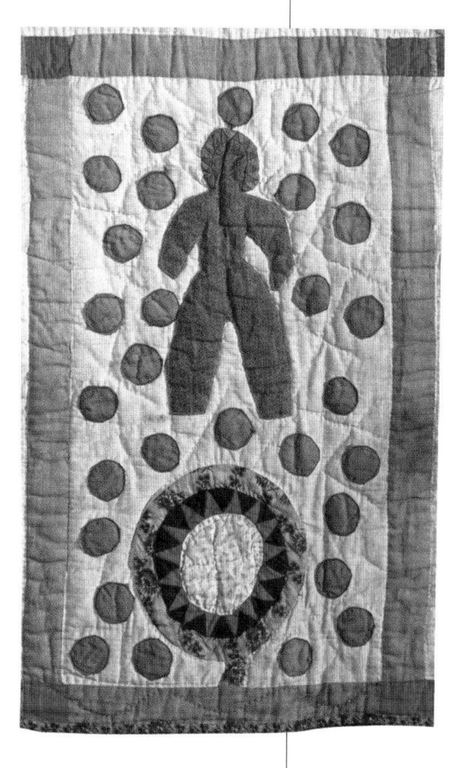

Section of the Bible quilt depicting Judas Iscariot standing among silver pieces and a star.

its maker. In colonial America, quilts were primarily a form of domestic artwork created by women of wealth who had sufficient resources and leisure time.

With the large-scale manufacture of textiles during the Industrial Revolution, women needed to spend less time spinning and weaving fabric for basic family needs. By the 1840s a wide variety of commercial fabrics were attainable for most American families. In 1856 you could buy a Singer sewing machine on a payment plan. More American women were quilting, and they were inventing styles that took advantage of the newfound diversity of fabrics. Quilting also got political—becoming, for example, a significant source of fundraising for the abolitionist movement; some were even stitched with antislavery poems.

Harriet Powers was born on October 29, 1837, in Madison County, Georgia. As she wrote later in a letter to a prospective quilt buyer, Powers spent her childhood working on a plantation owned by John and Nancy Lester and "commenced to learn at 11 years old" how to read and study the Bible on her own. She married Armstead Powers at the age of eighteen. She had nine children, but at least six died during her lifetime. After Harriet and Armstead became free at the end of the Civil War, they moved to Dondy, Georgia, and started a small farm.

It wasn't until 1886, once her children were grown, that Powers began exhibiting quilts at local fairs and competitions. They received immediate attention, winning prizes and bids from competing buyers. Powers refused to sell the Bible quilt for many years, however, only doing so when she was forced to by extreme financial hardship. The buyer, a white woman named Jennie Smith, described the scene in a handwritten note that is now preserved at the Smithsonian: "She arrived one afternoon . . . in an ox-cart, with the precious burden in her lap encased in a clean flour sack. She offered it for ten dollars, but I only had five to give. After going out and consulting her husband . . . she obeyed. After giving me a full description of each scene with great earnestness and deep piety she departed, but has been back several times to visit the darling offspring of her brain. She was only in a measure consoled for its loss when I promised to save her all my scraps."

If Powers was uniquely attached to this "darling offspring of her brain," it was perhaps because she had stitched so much of herself into it. Each of the Bible quilt's eleven panels are layered with meanings ranging from the cosmic to the deeply personal. In panel six, for example, Powers depicts Jacob's Ladder, a story from the Book of Genesis in which

Section of the Bible quilt depicting Adam and Eve naming the animals.

the biblical patriarch Jacob, hunted and homeless, discovers a ladder ascending to heaven. Jacob's Ladder had become a shared allegory of escape among American slaves, inspiring "We are Climbing Jacob's Ladder," one of the most popular spirituals of antebellum America.

In the Bible quilt, Powers drew not only from scripture but also science. Panel nine, depicting Judas Iscariot surrounded by thirty pieces of silver, contains a reference to a star that appeared in 1886 for the first time in 300 years. In a subsequent work, a pictorial quilt from 1898, Powers continued to intermix biblical scenes with cosmic wonders of her era, including the Leonid meteor shower of 1833, the Georgia cold front of 1895, and the infamous Dark Day of 1780, when New England mysteriously fell into an almost complete darkness. In this way, Powers's quilts resembled the scientific almanacs of her day. Like Benjamin Banneker before her, Powers recorded celestial phenomena, weather patterns, and snippets of scripture into the same ledger. Her quilts are a reminder that elements of religion and science can come together seamlessly within the lived experiences of American people.

Like other great works of art, the Bible quilt somehow seems to be composed of the materials and histories of its time while transcending them. In contrast to the precise geometric and floral quilts of its time, the Bible quilt brings together a menagerie of fabrics and figures within irregular blocks, set against a zigzagged backing of intersecting stitches. "Harriet was one among many quilters who insisted on doing things her own way," writes historian Laurel Thatcher Ulrich. Her "quilts are unlike any others not because they are unrelated to work being done around her but because she, like many other women of her generation, treasured innovation."

Engraved by John Sartain, Phila
JOHN WILLIAM DRAPER, M.D. LL.D.
PROFR OF CHEMISTRY AND PHYSIOLOGY IN THE UNIVERSITY OF NEW YORK

Conflict or Agreement?

The nineteenth century saw a major change in the way Americans understood the natures of science and religion. In early America, the line between religion and science did not seem so firm. Practitioners of scientific observation and invention like Cotton Mather and Benjamin Banneker held a wide range of religious beliefs, and Thomas Paine's vision at the orrery—of a mechanical explanation for all things—was still considered radical. After Darwin, however, Americans increasingly saw science and religion as two distinct spheres, and the explanatory power of the former seemed unlimited.

Two dueling notions emerged during this period, both premised on the idea that religion and science were fundamentally separate entities. One notion was that religion and science are essentially in conflict. The "conflict thesis" held that intellectual progress had for centuries been stymied by traditional mythologies and power structures; religion, in this view, was a vestige of human culture that science would eventually overcome. A second notion, which also took root in America during the nineteenth century, held that religion and science could ultimately achieve perfect harmony. These two notions—that science and religion must either inevitably clash or work in concert—are by now so commonplace that it seems surprising that each have a distinct history.

The conflict thesis emerged first and, depending on one's position on the matter, the physician, chemist, and photographer John William Draper (1811–1882) deserves a good portion of either the credit or the blame. When Draper was named the first president of the American Chemical Society in 1874, a fellow chemist declared that "few men of science now living in America have been so long and so favorably known." That same year Draper published his *History of the Conflict between Religion and Science*, which ambitiously framed the entire development

Philosopher, physician, chemist, and photographer John William Draper, ca. 1864.

of Western Civilization as a building toward an unavoidable showdown between irreconcilable ways of viewing the world. Though the book itself is somewhat dated, its so-called "conflict thesis" came to dominate discussion of science and religion during the century and a half that followed its publication, and it continues to have popular resonance today.

"The history of Science is not a mere record of isolated discoveries," Draper wrote in the influential text. "It is a narrative of the conflict of two contending powers, the expansive force of the human intellect on one side, and the compression arising from traditionary faith and human interests on the other." This conflict, he insisted, did not occur in a vacuum but rather is the inevitable outcome of two sources of ultimate authority vying for influence within the same culture. "The antagonism we thus witness between Religion and Science is the continuation of a struggle that commenced when Christianity began to attain political power," he argued. "A divine revelation must necessarily be intolerant of contradiction. . . . But our opinions on every subject are continually liable to modification, from the irresistible advance of human knowledge."

Forged in the anti-Catholic and anti-immigrant politics widespread in the middle of the nineteenth century, the "conflict thesis" would long endure in the American imagination.

As a man of science rather than an iconoclast or revolutionary, Draper presented himself as a somewhat reluctant truth teller—one who recognized that society had long sought to preserve peace by looking the other way when faced with religion's unearned claims. However, as the nineteenth century progressed, with ever more dire stakes in the political turmoil both in America and Europe, he felt called upon to speak.

Much of Draper's argument depended on narrow definitions of religion and science. Faith, he argued, was "unchangeable, stationary," while "Science is in its nature progressive; and eventually a divergence between them, impossible to conceal, must take place. It then becomes the duty of those whose lives have made them familiar with both modes of thought . . . to compare the antagonistic pretensions calmly, impartially, philosophically. History shows that, if this be not done, social misfortunes, disastrous and enduring, will ensue."

Rather than contending with religion in general, Draper focused his argument against the Roman Catholic Church, establishing a dichotomy of Protestantism as a forward-looking religious sensibility open to inquiry and innovation versus Catholicism as a relic of the Dark Ages. The true tension of his conflict thesis was thus between science and Rome: "Science . . . has never sought to ally herself to civil power. . . . She presents herself unstained by cruelties and crimes. But in the Vatican—we have only to recall the Inquisition—the hands that are now raised in appeals to the Most Merciful are crimsoned. They have been steeped in blood!"

Oldest surviving photograph of the moon, taken by John William Draper, 1840. Using his telescope and a photographic method known as the daguerreotype, Draper captured the moon in vivid detail.

Forged in the anti-Catholic and anti-immigrant politics widespread in the middle of the nineteenth century, this was a framework that would long endure in the American imagination. Echoes of Draper's rhetoric could be heard in popular anti-Catholic diatribes even in the middle of the twentieth century, such as 1949's bestselling *American Freedom and Catholic Power*, which warned that the Roman church was opposed to "the American gospel of science."

While undoubtedly influential, Draper alone was not responsible for the rise of the conflict thesis. Shortly before his *History* was published, Andrew Dickson White (1832–1918), a historian and cofounder of Cornell University, began to lecture on the "Battle-fields of Science" and brought forth his own summation of the relationship between scientific and religious ideas in the periodical *Popular Science Monthly*. "In all modern history," he wrote, "interference with science in the supposed interest of religion ... has resulted in the direst evils both to religion and to science. All untrammeled scientific investigation, no matter how dangerous to religion some of its stages may have seemed ... has invariably resulted in the highest good of Religion and of Science."

The title of a book White published twenty years later elaborating on this thesis, *A History of the Warfare of Science with Theology in Christendom*, signaled a significant escalation. As the nineteenth century gave way to the twentieth, the relationship of science and religion was, in the eyes of some Americans, no longer marked by mere conflict; now it would be all-out war.

Mary Baker Eddy, 1916. Eddy sought to reform Christianity through a new science of healing.

Forceful though Draper and White's arguments may have been, their declarations about the inevitable "warfare" between religion and science did not prove accurate. As contemporary historians argue, the conflict thesis depends on overly narrow conceptions of both science and religion and ignores the many documented examples of religion and science interacting in close alliance. "As a historical tool," argues historian Colin Russell, "the conflict thesis is so blunt that it is more damaging than serviceable."

Even as the likes of Draper and White were appraising the incompatibility of science and religion, others were seeking new areas of overlap and mutual influence. It is an intriguing historical coincidence that the year the conflict thesis was offered to the world also brought forth perhaps the most significant attempt to establish harmony. One response to the growing sense of controversy between science and religion was the effort by some religious groups to claim "science" as their own. Foremost among them was Mary Baker Eddy (1821–1910) and her Christian Science movement.

Born in 1821 in Bow, New Hampshire, Mary Baker Eddy was an author, religious leader, and a pioneering thinker who combined spirituality and health into a new religious movement. In 1879 she founded the First Church of Christ, Scientist, in Boston. Eddy's new church sought to "reinstate primitive Christianity and its lost element of healing" by establishing that sickness is ultimately an illusion that could be corrected by prayer. Between 1890 and 1926, Christian Science was the fastest growing religion in the United States. Its official newspaper, the *Christian Science Monitor*, would go on to win seven Pulitzer Prizes.

"If Christianity is not Scientific, and Science is not Christian, then there is no invariable rule of right."

First published in 1874 (the same year as Draper's *History*), *Science and Health with Key to the Scriptures* was Eddy's textbook for a new sect that quickly attained national significance. Her early influences included many of the religious currents growing in popularity and interest in the latter half of the nineteenth century, including Spiritualism, New Thought, and even Hinduism, which had begun to be introduced to the American public through Transcendentalist writers such as Ralph Waldo Emerson and Henry David Thoreau.

Eddy developed her theories on spiritual healing after her own struggles with injury, poor health, and personal loss. "Health is not a condition of matter, but of Mind," she claimed. Having found a "mind-cure" for herself through her own interpretation of scripture, Eddy used it to build a church and a publishing empire.

In considering the relationship between religion and science, Eddy had little patience for extreme views, which she deemed hypocritical. How will Christians follow the command of Jesus to preach the Gospel and heal the sick, she argued, if they ridicule science instead of integrating it into their practices? Eddy believed that if Christianity and science were not fused, it would ultimately lead to moral relativism. "If Christianity is not Scientific, and Science is not Christian, then there is no invariable rule of right," she argued, "and Truth becomes an accident."

Christian Science was neither the first nor last religion to embrace some form of science in America. During the nineteenth century, practitioners from a variety of religions and spiritual practices—including Mormonism, Buddhism, Judaism, Islam, and diverse Indigenous traditions—articulated their own visions of how science could be harmoniously integrated into their practices and worldviews. Much to the chagrin of those drawing battle lines between science and religion, these equally persuasive voices were seeking to erase the boundaries altogether.

COM
PLEX
ITY

Religion, science, and technology continued their entwined American history in the twentieth century. Despite efforts to reinterpret religious teachings to be more in accord with such new understandings as the age of the Earth, dramatic events including the Scopes trial highlighted tensions between science and some religious worldviews. Belief in the inevitable conflict between religion and science became widespread, even as religious ideas continued to provide inspiration to individual scientists, and the revelations of the past offered evocative language for describing discoveries that challenge human understanding.

Within this landscape, the 1965 revision of US immigration policy known as the Hart-Celler Act changed both American science and American religion dramatically. Throughout the 1970s and 1980s immigrants from South Asia, East Asia, and the Middle East had an outsize influence on medicine, technology, and related industries in the United States. Just as mid-century Jewish immigrant scientists played an important role in the breakthroughs of the 1950s and 1960s, today new groups of religious minorities are leaving their mark. Less than 1 percent of Americans practice Hinduism, for example, but Hindu Americans make up more than 5 percent of those working in technology, engineering, programming, and research. Likewise, the numbers of American Muslims and Buddhists remain relatively small, but their influence plays a major role in the medical and technology sectors. As a nation of immigrants, America continues to be shaped by new religious and scientific ideas.

Outdoor proceedings of the Scopes trial in Dayton, Tennessee, 1925. The trial, which centered around the teaching of human evolution in public schools, drew national publicity.

Her Heavenly Radium

Polish-French physicist and chemist Marie Skłodowska Curie, 1901.

A turning point in human history that made possible many of the advances and the horrors of the twentieth century, the 1898 discovery of radium by Marie Curie (1867–1934) also had implications for religion—particularly in America, where churches and other religious communities have often made use of new scientific developments to bolster the relevance of faith.

Born in Poland as Marie Skłodowska, she arrived in Paris as a young woman and soon met fellow physicist Pierre Curie, whom she married and began working with in 1895. While her husband was further along in his career when they met, it was Marie Curie's research that soon set the path their lives would take.

While pursuing questions related to the recent discovery of radiation in uranium for her thesis project, Curie studied other elements to determine if they too emitted this previously unknown source of energy. "There must be, I thought, some unknown substance, very active, in these minerals," she noted at the time. Soon she was able to announce the discovery of two new elements: polonium, named for the homeland, and radium, for which she coined the phrase "radio-active."

Of her two discoveries, it was radium that captured the public imagination. It became a cultural sensation due to its otherworldly glow and was even believed to have miraculous healing properties. Radium was added to everything from clocks and toys to toothpaste and chocolate. Within just a few years of the Curies' announcement of their discovery, the word "radium" had moved from their Paris lab to American pulpits. Sermons on "heavenly" and "spiritual" radium sought to identify the power of the mysterious and powerful substance with the messages of the Bible. As a Missouri Baptist minister preached in 1904: "God is the perfect radium of the universe, for [The First Epistle of] John says, 'God is light and in him is no darkness at all.' . . . We are proud of our great nation, and we have a right to be, but these blessings are ours because of the love and the light of the 'heavenly radium.' . . . It is God's will that we live in the light."

Ten years later a pastor compared radium's alpha, beta, and gamma rays to the Trinity, while others wondered if the great value suddenly assigned to this new element might be a sign that a generous deity had left untold marvels buried in the Earth, like coins beneath a sofa cushion: "We are constantly being reminded by new research, of the treasure which God has implanted in the physical world. Electricity is ever rising into new and more perfected uses. . . . The experiments with a new element, radium, so costly that an ounce of it is said to worth 32 millions of dollars, suggest the thought that God may have hidden riches in store for his human creatures, such as we yet hardly conceive of." Such divine associations with radium can be found in US sermons and the press through the 1950s.

As had been the case with the telegraph, Christians were not the only ones to seek religious meaning in new discoveries. Referring to the prominent role played by women as mediums in the previous century's Spiritualist movement, one adherent wrote in 1909: "Radium is another revelation of God's to a woman to find out something more about this wonderful network of electrical currents and nerves through every living body of mankind of which this universe is also full of another kind of currents, not discovered yet by man."

Though known as atheists, the Curies were not entirely separate from this spiritual and unscientific use of their work. At the height of their fame, they fell into the circle of the Italian medium Eusapia Palladino, who, when she toured New York, was referred to as the "Despair of Science." Though they were the most famous scientists in the world at the time, the Curies attended Palladino's séances somewhat credulously as examinations of other uses of unseen forces. "I must admit that those spiritual phenomena intensely interest me," Pierre wrote to Marie. "I think in them are questions that deal with physics."

One lasting influence of Curie's work on religion can be seen in the recalculation of the age of the Earth that her experiments made possible. Throughout the twentieth century efforts were made to reconcile a radium-informed Earth age far older than religious accounts of the Earth's creation. In the wake of Marie Curie's American visit, the popular press sought to make sense of this new timescale, particularly as it related to traditional understandings derived from scripture. Under the headline "Why All Conflict Should Cease Between Science and Religion," a 1922 article in the *Brooklyn Daily Eagle* noted, "As to the age of the Earth scientists today differ greatly. None of them venture to assert its exact age to the very year, day of the month and even hour of the day. They

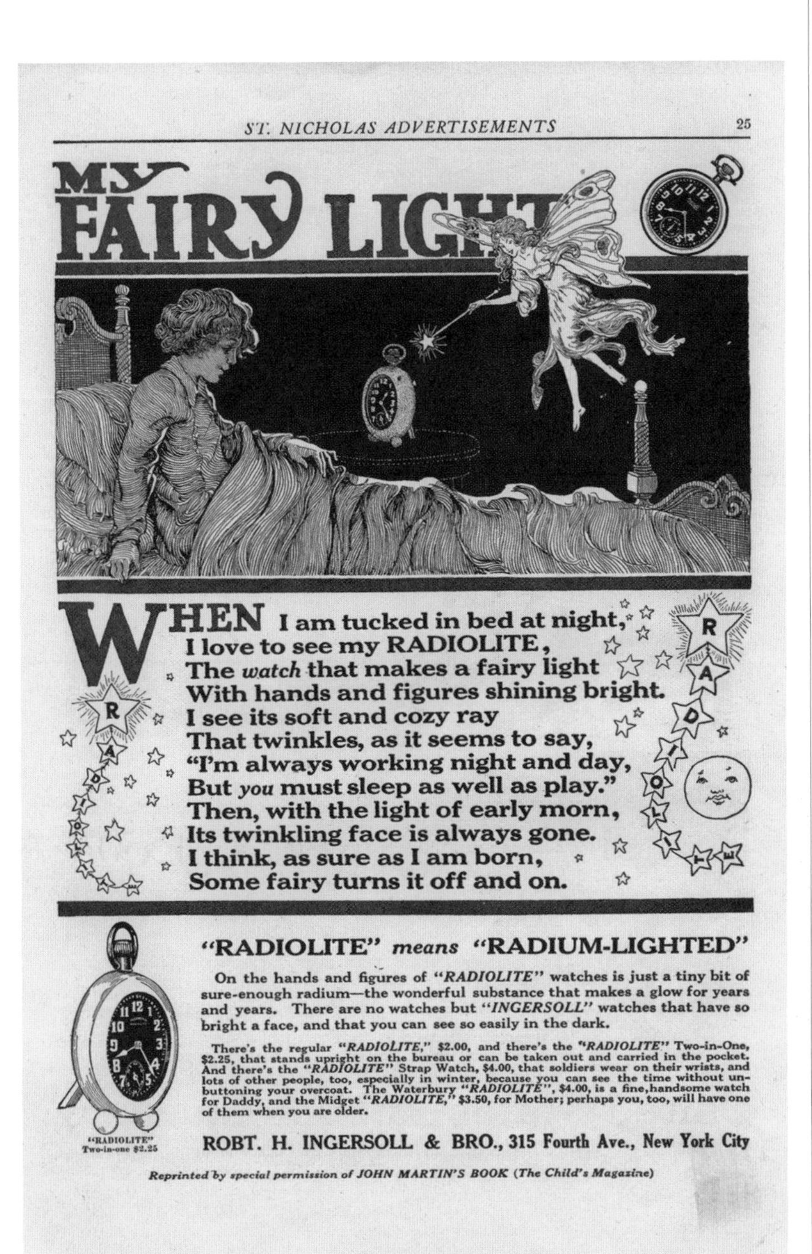

Ad for Ingersoll "Radiolite" clocks, ca. 1920. Before its negative health effects were understood, radium was added to consumer products to give them a miraculous glow.

are satisfied to place the age of the Earth approximately in millions of years. Some say it may be 100,000,000 years; some 200,000,000, and a French scholar has recently placed it at 500,000,000."

Yet none of this, the paper insisted, need be seen as a refutation of the stories of creation in which so many had invested their faith. On the contrary, discoveries like those made by Marie Curie merely expanded the grandeur of the universe. As the religious response to radium showed, there was no scientific revelation so disruptive it could not be put to the service of belief.

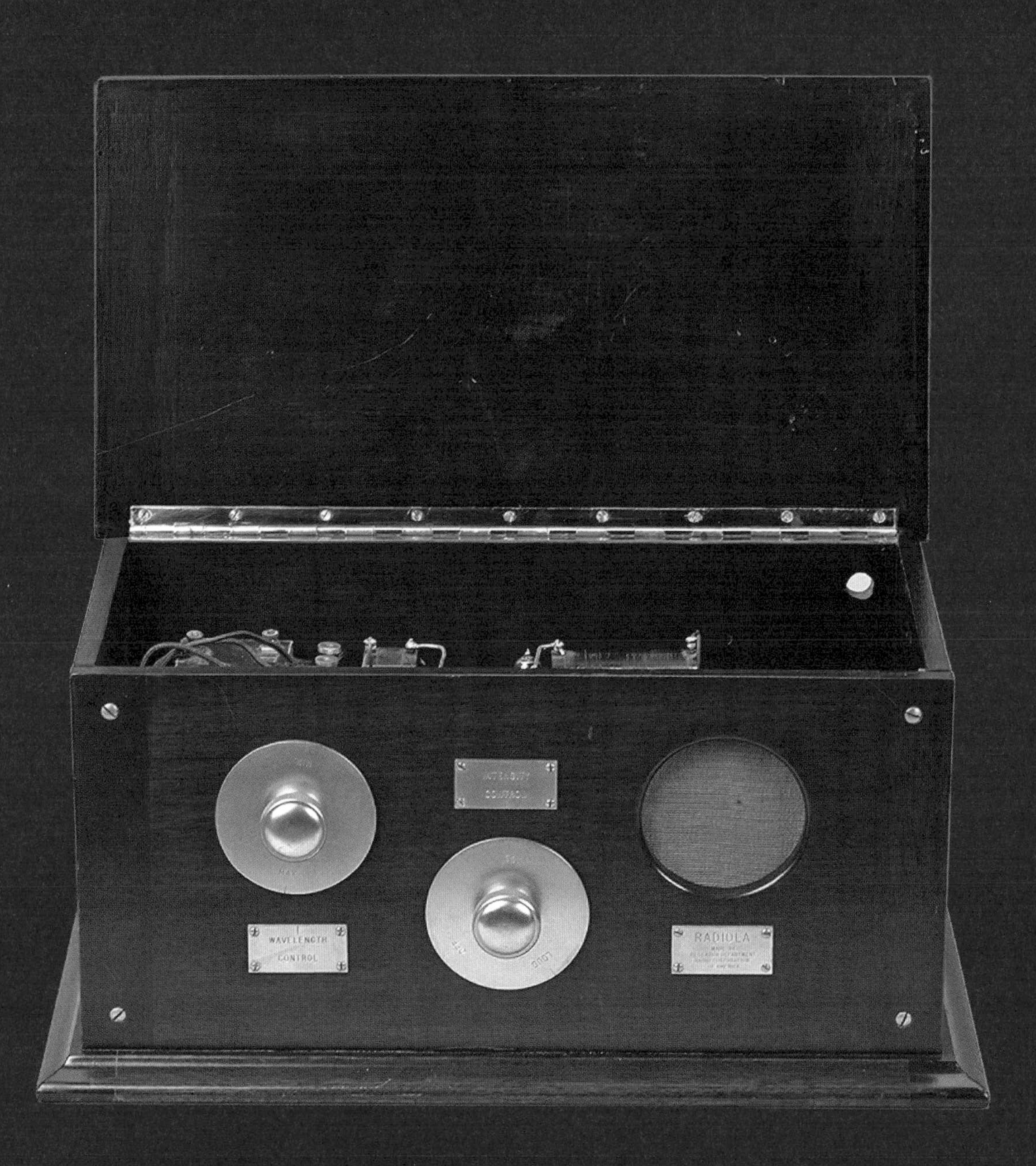
WAVELENGTH
CONTROL
RADIOLA

Religion on the Radio

RCA "Radiola I" prototype radio broadcast receiver, 1921.

For a fleeting hour one winter Sunday in 1921, a pious, infirm Massachusetts woman joined a religious community without leaving home. Her son had recently become a radio enthusiast, and though she had never imagined she would have much use for his wireless receiving set, he sat her down that evening and cupped two aluminum headphones over her ears. Four hundred miles away, in Pittsburgh's Calvary Episcopal Church, a congregation had just gathered for vespers in the presence of four microphones arrayed to capture the sounds of the pastor, the chimes, the organ, and the choir. Then, miraculously, she heard them.

"Last night for the first time in twenty years, I heard a full church service," the woman wrote the next day. "I could scarcely believe my ears when the organ music and the choir sounded distinctly. Then afterwards, the voice of the pastor thrilled me as few things have in the long suffering years. I kept the phones on all through the service and at the end felt at peace with the world." With a scriptural flourish, she added that this was no ordinary feeling of calm, but "the peace that passeth all understanding."

For those of us who, in the past thirty years or so, have come to spend our waking lives with eyes fixed on glowing screens, it may be difficult to remember that there was an earlier refocusing of our culture's collective attention that was no less revolutionary. Just as Samuel F. B. Morse's telegraph had accidentally transformed some elements of religion eight decades before, the introduction and popularization of the radio in the first quarter of the twentieth century did not at first seem to be religious in nature, but in retrospect it proved one of the greatest revelations the world had ever seen.

The nature of the change brought by radio was not as straightforward as it might at first seem, for it did not produce the immediate rejection of long-held lifeways that sometimes follows adoption of a new technology. Consider the context the Massachusetts woman provided for her initial encounter with radio: "for the first time in twenty years" she heard a full church service, some of which excited her "as few things have in the long suffering years." For her, radio was significant not only because it allowed her to experience something new, but because it allowed her to reexperience something old. Radio meant a recovery of a feeling lost but apparently longed for. The world-shrinking invention her son had installed in their home could have been brought from the future. Yet, in its ability to approximate the sensations of a bygone community within the confines of her isolation, radio for her was primarily a nostalgia machine. In this, at least, she was not alone.

Among the most popular programs in radio's first decades was Charles Fuller's *The Old Fashioned Revival Hour*. A Baptist preacher, Fuller became interested in radio in 1925, and within fifteen years he claimed his Sunday religious variety show had "the largest primetime distribution of any radio program in the country."

Much of the draw came from the use of "old time" hymns, which appealed not only to churchgoers but to those with perhaps more memories of Sunday services than histories of active participation. Letters sent to Fuller echoed the Massachusetts woman's yearning for an abandoned community, for the Proustian sounds of worship in the company of other

Radio transmission set used in the 1920s by the Church of the Covenant in Washington, DC, as seen in a 1938 exhibition at the Smithsonian Institution's Arts and Industries Building.

bodies: "We do enjoy the broadcasts and can hardly wait from one Sunday to the next. Our radio battery was down this last Sunday night, so we drove six miles over to another old couple's house to hear your sermon and that wonderful singing, and listened altogether to the songs we used to hear when we were children."

Notes of thanks for the reconnection Fuller provided to the remembered religious past came from those disconnected by age, illness, and geography: "I cook on a ranch out in the country and never get to go to church. I have not been to a church in nine years. God has blessed my little radio and through it the message has saved me. I am so hungry to hear more I can't wait from one Sunday to the next."

"God has blessed my little radio and through it the message has saved me."

Providing a glimpse into the spiritual lives of Americans during the interwar period, letters such as these also highlight a great irony hidden in the success of religious broadcasting: Among parts of the rural population in the United States, wireless receiver-set ownership rose as church membership fell. While not necessarily a direct cause and effect, this inverse relationship does raise a question that cuts to the heart of the influence of technology upon belief: If radio became "church" for those who had left actual churches behind, how did that affect the content of the faith?

The receiver set shown at the opening of this essay, the RCA Radiola, may not seem a religious object, but for many beginning in the 1920s that's just what the household radio became. A prototype receiver made by Alfred Goldsmith in the same year that the first church services were broadcast, it was designed to simplify the process of listening and so featured just a few basic controls: one to turn the unit on and off and change the volume, another to select the station. The ability to silence a sermon or to wander through the airwaves in search of different spiritual offerings no doubt had implications for personal devotion that perhaps we will never know. Religion on the radio had particular significance in the African American community, for whom hearing Black preachers on the airwaves also signaled the growing interest in related art forms, including gospel music, which through radio and records found a wide audience far beyond the walls of any church.

Much has been written about the ways religious organizations—and charismatic preachers, especially—seized on radio and made it their own, but relatively little has been said about those on the receiving end of faith-based transmissions or about the varieties of religious experiences created by radio professionals with no specific religious agenda. From the beginning, the spiritual significance of radio reached further than even the most ambitious of religious broadcasters could have hoped. Spreading well beyond the intentions of preacher broadcasters, radio did more than preach the Christian message to remote places; it wrote a gospel of its own.

Trial of the Century

Of the various controversies between religion and science in American history, the so-called Scopes Monkey Trial stands out as the most popular and, perhaps relatedly, the most emblematic of the conflict thesis. The trial set the stage for a notion that would persist throughout twentieth-century America: that the theory of evolution by natural selection is antithetical to Christian beliefs. With atheists and believers alike having turned Darwinism into a symbol of the intellectual tensions between science and religion, "Scopes" has stood as shorthand for this dispute for nearly a century.

The Scopes saga began in March 1925, when Tennessee signed into law a bill put forward by state congressman John Washington Butler that effectively outlawed the instruction of Darwinian theory. "It shall be unlawful," the Butler Act declared, "for any teacher in any of the Universities, Normals, and all other public schools of the State ... to teach any theory that denies the Story of the Divine Creation of man as taught in the Bible, and to teach instead that man has descended from a lower order of animals."

Later that spring, the five-year-old American Civil Liberties Union began to place newspaper notices throughout the state, seeking teachers willing to challenge the new anti-evolution law in the courts. Sensing an opportunity for press for their small town, a group of businessmen in Dayton, Tennessee, asked substitute high school teacher John T. Scopes (1900–1970) if he was willing to take part. As soon as Scopes agreed, events quickly took on a life of their own.

Despite being associated with the trial all the rest of his days and beyond, Scopes always insisted his role was minimal. He sometimes lamented the "fine mess" he had gotten himself into, noting that he had merely "furnished the body that was needed to sit in the defendant's chair."

This was not mere humility. Scopes was primarily a teacher of physics and math. While he accepted evolutionary theory, the extent

Tennessee schoolteacher John T. Scopes, 1921.

Succeeding pages: Politician and orator William Jennings Bryan (seated at left) spoke on behalf of Scopes' prosecution, while lawyer Clarence Darrow (standing at right), argued for the defense, July 20, 1925.

of his involvement in teaching about Darwin's ideas was limited to his use of a textbook: *A Civic Biology* was an overview of biology geared toward "beginners in secondary school" then widely in use across the state. In total, *A Civic Biology* devotes just two pages to the subject of evolution and barely more than a dozen words to Charles Darwin. Its surprising role as the flashpoint in a looming culture war can be found in a handful of sentences:

Evolution of Man. — Undoubtedly there once lived upon the earth races of men who were much lower in their mental organization than the present inhabitants. If we follow the early history of man upon the earth, we find that at first he must have been little better than one of the lower animals.

Broad though this passage was—note that it does not even say conclusively that early humans descended from lower animals, only that they were "little better" than them—Scopes was accused of violating the Butler Act.

This resulting trial was no ordinary legal procedure. Many of its key participants considered the proceedings to be symbolic of larger cultural

conflicts and social issues. Having inspired the case, the ACLU stepped in to finance Scopes's defense. The town of Dayton, meanwhile, played up the trial with the goal of attracting publicity—and found more than it could have ever dreamed when two of the best-known orators of the era, William Jennings Bryan (1860–1925) and Clarence Darrow (1857–1938), agreed to serve as attorneys for the opposing sides.

The country's preeminent and most acerbic journalist, H. L. Mencken (1880–1956), was on the scene throughout, providing commentary on the trial that would sear it in public memory. "The Scopes trial, from the start, has been carried on in a manner exactly fitted to the anti-evolution law and the simian imbecility under it," he wrote. "There hasn't been the slightest pretense to decorum. The rustic judge, a candidate for re-election, has postured before the yokels like a clown in a ten-cent side show, and almost every word he has uttered has been an undisguised appeal to their prejudices and superstitions."

From the start, popular culture also took an interest in the proceedings, as evident in the novelty records that soon provided an alternate take to Mencken's poison pen. Musician Vernon Dalhart recorded his own ode to Scopes in 1925:

Even the privies near the courthouse during the Scopes trial were involved in the debate between religion and science.

Oh, the folks in Tennessee
Are as faithful as can be,
And they know the Bible teaches what is right.
They believe in God above
And His great undying love
And they know they are protected by His might.
Then to Dayton came a man
With his new ideas so grand
And he said we came from monkeys long ago.
But in teaching his belief
Mr. Scopes found only grief
For they would not let their old religion go.

On the seventh day of the trial—with proceedings having moved outdoors due to excessive heat—the defense made the surprise move of calling prosecutor William Jennings Bryan himself to the witness stand. Darrow interrogated Bryan for two hours, questioning him on the book of Genesis, including the creation of Eve from Adam's rib, and arguing that the Bible was unscientific and therefore unfit for use in the classroom. Darrow eventually lost the trial—Scopes was found guilty and fined $100—but he was celebrated by evolutionists for this triumphant display. For his part, Bryan, too, was lionized by millions. The presidential contender and former secretary of state died in Dayton just days after the trial concluded, possibly due to the searing heat. Thousands visited his open casket, and crowds lined the railroad tracks as his body was transported to Arlington National Cemetery for burial.

While it is often caricatured as a debate between faith and reason, elements of the Scopes trial raised more significant questions than is popularly remembered. To begin with, the textbook at the center of the case, *A Civic Biology*, was far from an unbiased work promoting science education. Included in its two pages on evolution were passages that are shockingly racist, highlighting concerns over the social uses to which Darwinian theory has been put:

The Races of Man. — At the present time there exist upon the earth five races or varieties of man, each very different from the other in instincts, social customs, and, to an extent, in structure. These are the Ethiopian or negro type, originating in Africa; the Malay or brown race, from the islands of the Pacific; the American Indian; the Mongolian or yellow race, including the natives of China, Japan, and the Eskimos; and finally, the highest type of all, the Caucasians, represented by the civilized white inhabitants of Europe and America.

Elsewhere, disturbing sections on eugenics describe in detail human populations among whom, the book claims, "feeblemindedness," "immorality," and "parasitism" run rampant. As the text chillingly continues:

If such people were lower animals, we would probably kill them off to prevent them from spreading. Humanity will not allow this, but we do have the remedy of separating the sexes in asylums or other places and in various ways preventing intermarriage and the possibilities of perpetuating such a low and degenerate race.

While hardly in keeping with assumptions about the "right" side of the Scopes trial, such views undeniably were part of the approach to the teaching of evolution then being defended. As ever, the notion of the conflict between science and religion was far more complex than is often supposed.

Scopes himself, it seems, was at home with that complexity. "I don't know if I'm a Christian," he said, "but I believe there is a God." Yet when it came to doing his job as an educator, the matter could not have been more straightforward. "All of biology is basically the story of the evolution of matter and life," he said. "I was hired to teach science, and I went ahead and taught it."

Playing Dice with the Universe

Albert Einstein (1879–1955) was born in Germany, but like many scientists before him, he immigrated to the United States to escape religious persecution. The Nobel Prize–winning theoretical physicist was visiting the United States in 1933 when Adolf Hitler came to power. Einstein, a world-famous Jewish scientist and an outspoken critic of the Nazi party, was declared an enemy of the German state. He sought asylum and began working at Princeton University, eventually receiving American citizenship in 1940.

Einstein revolutionized humanity's understanding of the cosmos with his theory of relativity, which introduced an entirely new framework for comprehending space and time. Yet when he wasn't upending Newtonian physics, Einstein was also a prolific writer about topics ranging from religion to politics to the ecstatic rapture of scientific inquiry.

With his untamable hair and characteristic pipe, Einstein was a charismatic public figure known for making sage-like pronouncements on life, the universe—almost everything. When asked about the cosmos, he was quick to evoke religious language. Einstein famously rejected quantum theory, for example, by saying that "God does not play dice with the universe." He also said that, when judging a physical theory, he would ask himself "whether I would have made the Universe in that way had I been God."

Despite these invocations, however, Einstein did not believe in God—at least not in the sort of God who stands apart from the universe itself. He doubted that the laws of physics left any room for divine action, and he outright rejected the notion of a personal God with humanlike characteristics. Einstein was also deeply concerned with the problem of evil. During his lifetime, Einstein witnessed two world wars, genocide, and the atomic bomb. In public lectures and private letters, Einstein questioned how a benevolent, omnipotent God could allow for such

Theoretical physicist Albert Einstein at times used religious language in playful and enigmatic ways, ca. 1945.

suffering. He was at times blunt in his dismissal of conventional religious ideas. "The word God is for me nothing but the expression and product of human weaknesses," he once wrote. "The Bible a collection of venerable but still rather primitive legends. No interpretation, no matter how subtle, can [for me] change anything about this."

Einstein did not, however, believe that religion had no place in society. Like Joseph Priestley, a fellow immigrant-scientist who came before him, Einstein believed that religion had a purpose and that science could further that purpose by making religion more rational. Einstein argued that religion was originally born out of the human need to make sense of suffering, but that it has been evolving ever since. With Judaism and then Christianity, he argued, religion developed into a moral system that could improve social life. The next step for religion, Einstein claimed, would be to embrace science and thereby move beyond dogmas, creeds, central churches, even the concept of God.

Eventually, Einstein thought, religion would center around "cosmic religious feeling," a state in which the "individual feels ... the sublimity and marvelous order which reveal themselves both in nature and the world of thought." While experiencing cosmic religious feeling, he argued, a person feels imprisoned by their individual existence and "wants to experience the universe as a single significant whole."

Einstein believed that cosmic religious feeling is what inspired great religious leaders of the past, including the Buddha. He also thought it was at the heart of science, since scientific research requires the same devotion and rapturous awe. In his 1954 essay "Science and Religion," Einstein described a paradoxical relationship that science and religion share. On the one hand, he said, they often do seem to be "irreconcilable antagonists," with churches persecuting the "devotees" of science.

On the other hand, I maintain that the cosmic religious feeling is the strongest and noblest motive for scientific research.... Those whose acquaintance with scientific research is derived chiefly from its practical results easily develop a completely false notion of the mentality of [devoted scientists who] remain true to their purpose in spite of countless failures. It is cosmic religious feeling that gives a man such strength. A contemporary has said, not unjustly, that in this materialistic age of ours the serious scientific workers are the only profoundly religious people.

This conviction perhaps gives new meaning to Einstein's famous aphorism that "science without religion is lame, religion without science is blind." Science without religion is lame because it lacks cosmic religious feeling, that transformative devotional awe that lifts the individual out of their isolated perspective to see the universe in an

Albert Einstein's pipe, made before 1948.
The Nobel Prize-winning physicist was rarely seen without his briar pipe, now among the most popular objects in the National Museum of American History's Modern Physics collection.

entirely new way—perhaps even reenvisioning space and time itself. Religion without science is blind because any system of meaning can turn inward, away from reality and toward myopic dogmas that ultimately create more suffering.

Beyond his status as a preeminent symbol of science itself, Einstein may be just as significant as the most famous representative of a phenomenon that has remade and refreshed both the scientific and religious landscape in the United States again and again. Early in the twenty-first century studies completed by the University of Chicago showed that professionals in scientific and medical fields were dramatically more likely to represent religious traditions arriving in this country in great numbers only through relatively recent immigration. Physicians, for example, are twenty-six times more likely to be Hindu than the overall US population; they are six times more likely to be Buddhist, and five times more likely to Muslim.

Such demographic trends demonstrate the ways in which the sciences have often been, counterintuitively to some, among the most religiously diverse and welcoming spheres of American life. Like Priestley before him, Einstein found in the United States a place all his radical ideas—scientific and religious—could call home.

HOLY BIBLE

Immortal Life?

When Henrietta Lacks (1920–1951) was diagnosed with cancer, she was unaware that doctors had taken cell samples from the tumor in her cervix. Those cells were given to Dr. George Gey, a cancer researcher at Johns Hopkins University, who discovered that there was something unusual about them: Lacks's cells were immortal.

In recent years, this strange fact about tissues from one woman's body, the fraught ordeal of how they were taken from her, what became of her family, and who profited from the cells' use has become a story that has captured the imagination of millions. The subject of a bestselling book and HBO film starring Oprah Winfrey, Henrietta Lacks has now made a remarkable impact on American culture both through media representations and through her actual cellular remains. Her story is also one of the most moving recent examples of the many ways religion, science, and new technologies intersect in individual lives. Far from abstractions, religious and scientific ideas are the essence of human experience.

Human cell lines are crucial to biomedical research. They can be used to study everything from genetics to cancer, and they are particularly valuable for creating and testing medicine. The problem with human cells is that they die. A cell's DNA degrades with every division, due to replication errors and the incremental shortening of telomeres during the copying process. Eventually every cell line is unable to reproduce itself. This process—known as "apoptosis" or "cellular suicide"—is a healthy biological function in living creatures, but it was a frustrating limitation for twentieth-century scientists. Typical human cell lines survive in laboratory conditions for only a few days, making experiments with them slow, tedious, and expensive.

Gey observed that Lacks's cells—now known as HeLa—behaved differently. They grew unusually quickly, doubling every twenty-four

Henrietta Lacks (HeLa): The Mother of Modern Medicine **by Kadir Nelson, 2017.**

hours. Where most cell lines cannot survive in test tubes, HeLa cells flourished, sometimes even escaping to take over other tissue cultures. Finally, HeLa cells did not succumb to apoptosis. They rebuilt their own telomeres after every cell division and thereby circumvented the process of cellular aging. Gey had discovered the first immortal cell line.

Henrietta Lacks passed away on October 4, 1951, at the age of thirty-one. She was survived by her husband and five children. She was also survived by the cancer that took her life, which would live on indefinitely and save countless lives. The HeLa cell line was utilized by scientists working in disease research, genetics, and medical testing. It became the cornerstone of virology, as researchers learned they could infect HeLa cells with viruses and observe their complex interactions. HeLa was sent into space to study how cancer cells are affected by gravity. Perhaps most significantly, HeLa helped save millions of lives through the work of Jonas Salk, who used the cell line to create and test the polio vaccine.

This detail from Kadir Nelson's painting shows Henrietta Lacks holding her Bible—a cherished family heirloom that has helped make meaning of her sacrifice.

As Rebecca Skloot writes in *The Immortal Life of Henrietta Lacks*, "Her cells were part of research into the genes that cause cancer and those that suppress it; they helped develop drugs for treating herpes, leukemia, influenza, hemophilia, and Parkinson's disease; and they've been used to study lactose digestion, sexually transmitted diseases, appendicitis, human longevity, mosquito mating, and the negative cellular effects of working in sewers. Their chromosomes and proteins have been studied with such detail and precision that scientists know their every quirk. Like guinea pigs and mice, Henrietta's cells have become the standard laboratory workhorse."

By way of her cells, Henrietta Lacks has been one of the most significant figures in American history. Yet the circumstances of HeLa's origins have raised serious ethical questions about medical consent, race, and the rights that individuals have over their own bodies.

Lacks, an African American woman, never consented to being a tissue donor. We will never know if she would have chosen to have a part of her body live on forever to be endlessly probed, destroyed, and made new again. Lacks's children received no financial benefit and were not even made aware of their mother's cells for more than twenty years. The social history of HeLa can trace its lineage back to far darker experiments made on African Americans (such as the Tuskegee syphilis study, and the experimental surgeries of J. Marion Sims), which often drew on prejudicial, pseudoscientific understandings of race.

For their part, many of Lacks's surviving family members came to understand the HeLa cell line through a Christian lens, taking the news as a sign that Henrietta had been chosen by God to take on an immortal spiritual body. In *The Immortal Life of Henrietta Lacks*, Skloot describes a conversation with Gary Lacks, Henrietta's nephew. "Henrietta was chosen," he said, "and when the Lord chooses an angel to do his work, you never know what they going to come back looking like." Gary then pointed in his Bible to John 10:28, a passage where Jesus makes a promise to his followers: "I give them eternal life, and they shall never die."

"Henrietta was chosen, and when the Lord chooses an angel to do his work, you never know what they going to come back looking like."

Gary Lacks's response is an example of how conflicts between science and religion are not always between progressive truth seekers and regressive traditionalists; in some circumstances, religion can provide a moral and intellectual framework for understanding something as uncanny and inspiring as a family member's immortal flesh.

Artist Kadir Nelson's portrait of Henrietta Lacks shown at the opening of this essay recently joined the collections of the Smithsonian Institution's National Portrait Gallery and National Museum of African American History and Culture. It seeks to present all the contradictions and layers of meaning present in the story of the woman who inspired it. The painting's background first appears to be an elaborate wallpaper but on closer inspection is an overlapping "flower of life" pattern, representative of human cells. She holds before her a Bible, positioned over the area of her body from which her cells were taken. Her head is ringed in a halo, suggestive of the angelic presence her family believes she remains in many people's lives.

"Nelson wanted to create a portrait that told the story of her life," National Portrait Gallery curator Dorothy Moss said of the painting. "It will spark a conversation about people who have made a significant impact on science yet have been left out of history."

mankind focused on a single event, and perhaps with total unanimity been prayerfully with three far-away men.

After you have addressed yourself to these areas, (and I would save some of it for the second telecast on the ninth orbit,) about the only thing I can think of to match the majesty of the occasion, and the evening, is to read the opening lines of Genesis. These lines are Christian the world over in the very real sense of the word, and I think would sound the universal appeal and sense of reverence that is called for.

You would be reading them while looking up at the Earth from the moon. You could switch to them by saying something like, "I would now like to read you the opening sentences of the Holy Scripture."

1. In the beginning God created the heaven and the earth.

2. And the earth was without form, and void; and darkness was upon the face of the deep. And the Spirit of God moved upon the face of the waters.

3. And God said, Let there be light: and there was light.

4. And God saw the light, that it was good: and God divided the light from the darkness.

5. And God called the light Day, and the darkness he called Night. And the evening and the morning were the first day.

6. And God said, Let there be a firmament in the midst of the waters, and let it divide the waters from the waters.

7. And God made the firmament, and divided the waters which were under the firmament from the waters which were above the firmament: and it was so.

8. And God called the firmament Heaven. And the evening and the morning were the second day.

9. And God said, Let the waters under the heaven be gathered together unto one place, and let the dry land appear: and it was so.

10. And God called the dry land Earth; and the gathering together of the waters called he Seas: and God saw that it was good.

I would THEN close with, "Good night, good luck, a Merry Christmas, and God bless you all--all of you on the good Earth." That ends the broadcast.

Earthrise

When the Apollo 8 mission orbited the moon on December 24, 1968, the three astronauts on board—Bill Anders (1933), Frank Borman (1928), and Jim Lovell (1928)—took turns reading the opening words of the Book of Genesis, which were then broadcast to the largest television audience in history. It was a moment that captured a key factor in the relationship between science and religion in American life: The space program would not have been possible if the worldviews of those involved had been limited to the kinds of knowledge religion provides, and yet in the midst of an epochal scientific and technological achievement, words from an ancient religious tradition were used to frame its significance, serving as a reminder at once of the boundlessness of human ambition and the limitations of human understanding.

It had been a year in need of uplift. Heartbreak and terror had greeted every season, beginning with the Tet Offensive in Vietnam in January, a series of attacks on South Vietnamese and US forces that eroded American public support for the war. This was followed by the assassinations of Martin Luther King Jr. and Robert Kennedy in April and June. By December the country was ready to turn its attention to the latest Apollo mission in hopes of vicarious escape.

From the start, the Apollo program had been as much about harnessing public attention as it had been about the scientific benefits of entering the moon's orbit. When the launch schedule made it clear that the astronauts would be in the vicinity of the moon at the height of the winter holidays, NASA knew something special ought to be planned.

Tasked only with saying "something appropriate" on Christmas Eve, the crew, in consultation with NASA advisers, hit upon the idea of reading aloud the opening verses of Genesis, the biblical account of the creation of the world. As a text drawn from the Hebrew Bible—revered by both Jews and Christians and respected by Muslims as a precursor to

Apollo 8 flight plan showing the live television broadcast script for Christmas Eve 1968.

the Qur'an—the passage was regarded as more appropriately inclusive than verses specifically related to the holiday would be.

As Borman said in 2008: "We were told that on Christmas Eve we would have the largest audience that had ever listened to a human voice, and the only instructions that we got from NASA was to do something appropriate." Lovell further explained, "The first ten verses of Genesis is the foundation of many of the world's religions, not just the Christian religion. There are more people in other religions than the Christian religion around the world, and so this would be appropriate to that and so that's how it came to pass."

Crowded in their capsule, Anders, Borman, and Lovell each read verses aloud from a typescript on fireproof paper that had been inserted in their flight manual, as seen at the start of this essay. It was the first time printed words of a religious tradition had traveled into space.

The astronauts knew the audience for their reading would be vast, but no one was quite prepared for the response they received. As National Air and Space Museum curator Teasel Muir-Harmony notes, "Around the world, television sets glowed with the broadcast. One in four people on Earth—roughly a billion people spread among sixty-four countries—listened to the reading." After their return to Earth, the crew received more than 100,000 letters in response.

Not everyone was pleased with government resources being used for a Bible broadcast, however. The atheist activist Madalyn Murray O'Hair (1919–1995) and her Society of Separationists brought a lawsuit

***Earthrise*, 1968. Taken from lunar orbit, this photograph of the Earth transformed the way people thought about their place in the cosmos.**

against NASA and its administrator, Thomas O. Paine, "seeking an order enjoining NASA from doing any act whatsoever which abridges the plaintiffs' freedom from religion or establishes Christianity as the official religion of the United States," which, O'Hair argued, reading the Book of Genesis from a government-owned space capsule seemed to do.

Apollo 8 astronauts Bill Anders, Jim Lovell, and Frank Borman, 1968.
As crew of the first piloted spaceflight to orbit the moon, they were the first humans to ever witness an "Earthrise."

"The various plaintiffs are atheists, deists, and believers in the complete separation of church and state," the court filings noted, who asserted their right to bring suit based on their status as taxpayers and citizens. They argued that their First Amendment rights had been hindered both by the broadcast itself and the decision to launch the Apollo mission during a religious holiday, which was bound to blur the lines between church and state in the actions of the crew and the perceptions of the public.

In dismissing the case, the judge noted that no one forced O'Hair to listen to the broadcast, and then added forcefully, "With regard to the scheduling of the Apollo 8 flight during the Christmas season, it is approaching the absurd to say that this is a violation of the Establishment Clause because of the religious significance of that date. The First Amendment does not require the State to be hostile to religion, but only neutral."

While the celestial Bible reading perhaps proved more controversial than NASA had hoped, another enduring symbol from Apollo 8 proved to be universally popular. As the astronauts began their fourth orbit of the moon, they caught sight of the Earth coming up on the lunar horizon. Working quickly, they were able to take the iconic photograph that would come to be known as *Earthrise*, which secured Apollo 8's place in history by providing humanity with a new way of seeing itself.

More than half a century later, *Earthrise* and the larger context of the mission during which it was captured serve as a dramatic illustration that the terms of the religion and science discussion often rest on appeals to two kinds of authority: the authority of spiritual traditions on the one hand, and the authority of observation, experiment, and the scientific method on the other. While at times simply in conflict, each of these sources of authority has also drawn on the other for legitimacy throughout American history. Practitioners of each have variously defined themselves in opposition to the other, borrowed the language of the other for purposes that expand its meaning, and, as with Apollo 8, discovered surprising moments of complementarity.

Your Brain on God

Americans have long been fascinated with the brain. During the nineteenth century, phrenology took the nation by storm with its notion that the bumps of a person's skull corresponded to their underlying personality traits. Phrenological busts mapped cognitive regions for everything from reason to spirituality, promising that even the most ephemeral parts of ourselves could be objectively measured. Even after phrenology was rightly abandoned as a science, it remained popular for decades as a method of self-improvement. If the brain could be mapped, it stood to reason, it could also be explored.

A century later, a new set of brain sciences emerged, as cognitive science, neuropsychology, and neuroimaging began to map the inner workings of the mind. President George H. W. Bush proclaimed that the 1990s would be the "Decade of the Brain." "The human brain, a 3-pound mass of interwoven nerve cells that controls our activity," he marveled, "is one of the most magnificent—and mysterious—wonders of creation." In the ensuing years, some of the most significant insights into the brain would come from dialogue between religious and scientific communities.

For the past three decades neuroscientists have used brain-imaging technology to study the brains of Buddhist monks and nuns, assessing the physical effects of religious practice. Led by Dr. Richard Davidson, the Center for the Healthy Minds at the University of Wisconsin-Madison has been a pioneer in this effort. Building on a 1992 meeting with the Dalai Lama, the spiritual leader of Tibetan Buddhists, Richardson and his team invited monks to meditate in a most unlikely setting: a scientific laboratory, where they donned a crown of electrodes that would measure the electrical activity of their brains. Using electroencephalography (EEG) and, later, functional magnetic resonance imaging (fMRI), Davidson observed the neural patterns of meditation in real time. What he saw shocked him.

Dru-gu Choegyal Rinpoche, a Buddhist teacher, wearing an array of sensors used for making electroencephalographs, as part of a study on the health effects of meditation at the Center for Healthy Minds, Madison, Wisconsin, 2003.

"This had never been seen in a human brain before," he said. In experienced meditators, it turned out, brains displayed not bursts of activity common in untrained minds but long durations—a lingering cloud rather than a puff of smoke that disappears in the wind. Homing in on the left prefrontal cortex, Davidson and his team found that meditation has long-term effects on the functions and structure of the brain. Subsequent dialogue between neuroscientists and Buddhist meditators have produced insights into the neuroscience of attention, emotion, and conceptual processing.

Around the same time, a decades-long legal battle over religious freedom and the use of mind-altering substances was coming to a resolution. In 1994 an amendment to the American Indian Religious Freedom Act provided that "the use, possession, or transportation of peyote by an Indian for bona fide traditional ceremonial purposes in connection with the practice of a traditional Indian religion is lawful, and shall not be prohibited." This overruled previous laws that did not protect the use of peyote, a species of cactus containing the psychedelic chemical mescaline, within the scope of the free exercise of religion.

Ceramic tray for serving peyote made by Rev. Immanuel P. Trujillo and used by the San Carlos Apache, ca. 1955. Long used by Indigenous peoples of the southern Plains in ritual practices, the hallucinogenic cactus has recently been studied for clinical treatments.

During its protracted legal struggle, the Native American Church, an intertribal organization representing tribes across the United States, found support from members of the scientific community. The journal of the American Association for the Advancement of Science published a statement in 1951 that defended peyote as having "remarkable mental and physical effects" with observable medical and spiritual value. "A scientific interpretation might be that the chemicals in peyote diminish extraneous internal and external sensations, thus permitting the individual to concentrate his attention on his ideas of God."

Subsequent scientific research has further validated the benefits of peyote and other psychedelic plants that have been used ceremonially by Indigenous peoples. Studies show that the active compounds in peyote, psilocybin, ayahuasca, and morning glory bind to serotonin receptors in the brain and thereby increase serotonin levels, much like selective serotonin reuptake inhibitors used to treat depression. In clinical research settings, this class of sacred plants has proven effective in treatment for addiction, depression, and anxiety, with participants reporting both spiritual experiences and substantive therapeutic outcomes. Psychedelics have recently been studied for the treatment of posttraumatic stress disorder among US Special Operations Forces veterans. This shift in peyote's classification speaks to a larger trend of recognizing Indigenous knowledge systems as a resource for understanding the natural world and the workings of the human mind.

The Decade of the Brain was not only a time when religious knowledge influenced the science of the mind, but also a period when neuroscience offered its own form of spiritual experience. As discoveries about religion and the brain involved ever more complex technologies, it did not take long for some to wonder if the brain might be hacked.

The laboratory apparatus popularly known as the God Helmet was developed by inventor Stanley Koren and neuroscientist Michael Persinger of Laurentian University. Made from a snowmobile helmet and some strategically placed solenoids, the device was designed to disrupt communication between the left and right temporal lobes using a weak magnetic field. A trained psychologist, Persinger hypothesized that when a person's temporal lobes are disrupted it creates the sense of an outside, visiting consciousness. This, he reasoned, could be the neural underpinning of various religious experiences, like feeling the presence of God or receiving a visitation from an angel, spirit, or UFO.

Developer of the "God Helmet," Dr. Michael Persinger, holding a preserved human brain, 1999. His neuropsychological studies attempted to create the sense of an external presence—the feeling that another person or entity is near—through weak magnetic fields.

Persinger and his colleagues measured the effectiveness of the God Helmet in a series of experiments. Test subjects wore the helmet and "opaque goggles while sitting quietly within an acoustic chamber that was weakly illuminated by a red light." There they received a succession of magnetic pulses aimed at their temporal lobes. Surveyed afterward, a majority claimed that they felt an abstract "presence" during the session. Some even reported mystical experiences and altered states of consciousness.

Persinger's work raised tantalizing questions, and the God Helmet quickly became a media sensation. Journalists declared that Persinger had discovered the so-called "God Spot" of the brain. The famous British scientist and outspoken atheist Richard Dawkins even visited Persinger's lab to try out the God Helmet himself.

In the thirty years since President Bush's proclamation, the brain has remained a "magnificent and mysterious" source of insight, for scientists and religious practitioners alike. It has facilitated, in the study of meditation, new exchanges between science and religion. From God Helmet hackers to Buddhist brain scans, the brain has inspired new ideas about what religion might be.

Beginner
Advanced
IPRAYER
Charged Charging Low
Vol -
Vol +
Fajr
Sunnah
Fard
Zuhr
Sunnah
Fard
Sunnah
Nafl
Asr
Sunnah
Fard
Maghrib
Fard
Sunnah
Nafl
Isha
Sunnah
Fard
Sunnah
Nafl
Witr
Nafl
Jumah
Sunnah
Fard
Sunnah
Sunnah
Nafl

The Technology of Prayer

Prayer has traditionally been defined as a ritual act of communication with a sacred entity. Whether directed toward a God, gods, spirits, ancestors, or a transcendent realm, prayer seems to be a human practice that spans many cultures. In his 1902 book *The Varieties of Religious Experience*, the American philosopher William James (1842–1910) went so far as to argue that prayer is "the very soul and essence of religion."

Writing during a time when science and technology seemed to be encroaching on the domain of religion, James saw prayer as an example of something that science could not fully capture. Prayer is "no vain exercise of words, no mere repetition of certain sacred formulae," he argued, "but the very movement itself of the soul." In other words, an automaton of a monk does not pray.

Yet, in the century after William James drew a firm line between prayer and technology, the two would only continue to intermix. In recent years, this has perhaps been seen most clearly in the explosion of what some have called the mindfulness industry, which has generated countless apps and gadgets designed to foster improved focus and well-being. A headset developed by the San Francisco tech company Emotiv, for example, may represent the next stage in the evolution of communication between humans and machines. Harnessing users' attention to control elements of the outside world simply by thinking, the technology has been compared to "The Force," but it is intended mainly as a method of training the mind. "You're doing the same thing as a meditator, a Buddhist monk might do," an Emotiv developer has said. "But maybe we, in the West, need a device to do it."

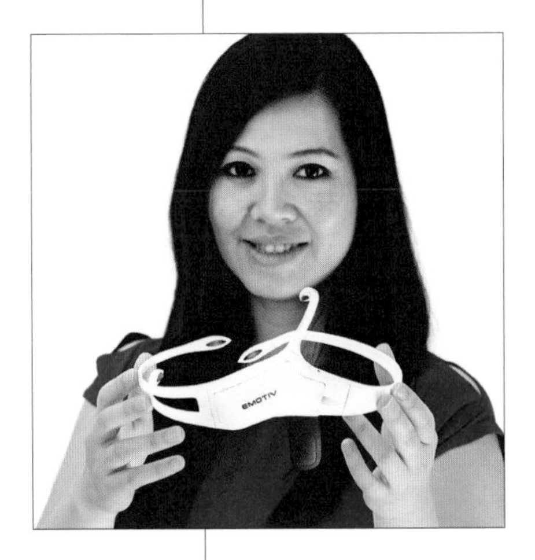

Tan Le, cofounder of Emotiv Systems, with a headset designed to allow users to focus mental abilities while gathering and analyzing brain data.

Opposite: "Islamic iPrayer" rug, 2014, uses digital audio to share ancient rituals.

The use of technology for spiritual training is influencing traditional religious communities as well. Modern Muslims, for example, have invented and utilized a variety of devices that combine technology and prayer in unexpected ways. This is no more apparent than in some of the machines that American Muslims use to perform daily prayer, or salat.

Salat is a foundational practice for Muslims across the globe. It is one of the five "pillars" of Islam, along with a profession of faith (shahada), charity (zakat), fasting (sawm), and a pilgrimage to Mecca (hajj). None of these practices are universal among Muslims—Islam, like other religions, is as diverse as the individuals who practice it—but salat is nonetheless a deeply significant ritual for Muslims the world over.

Salat is traditionally performed five times a day, beginning at dawn, at prescribed times based on the movement of the sun. Since the timing of prayers is based on celestial motion, they change based on the precise time and coordinates of the person praying. Muslims pray toward the Kaaba, a sacred site in Mecca, Saudi Arabia, that the Qur'an describes as the first House of Worship for all of humankind. Salat involves repeated cycles of physical movement and scriptural recitation in patterns that vary slightly for each daily prayer. Within these structures, salat is understood as a transformative connection to Allah that helps Muslims sustain their mental, emotional, and spiritual health.

According to the Pew Research Center, there were roughly 1.8 billion Muslims in the world as of 2015, with over 3 million living in the United States. What began in the seventh century as a small religious movement on the Arabian Peninsula is now the world's second-largest faith tradition. As Islam spread across the globe, so too did new techniques

With prayers directed toward the Saudi Arabian city of Mecca, Islam has long made use of location-finding technologies, such as this portable brass sundial and qibla indicator, ca. eighteenth century.

for performing salat. Some scholars argue that the technical exigencies of Islamic prayer—the need to calculate the direction of the Kaaba from any location, for example, or the importance of predicting the precise motion of the sun—helped spur important scientific breakthroughs in astronomy and mathematics. Innovation rarely happens in a vacuum; it is often a response to deeply felt human needs.

Technologies of salat proliferated during the twentieth and twenty-first centuries, adapting to an era of global migration and digital tech. Muslim inventors and engineers crafted devices like digital *azaan* clocks, which display each day's precise times for prayer, and multilingual electronic Qur'ans. Advances in location technologies like GPS made possible a new generation of tools for calculating the direction of prayer, or qibla. Google's dedicated "Qibla Finder" app advertises that users can "Locate the Qibla, wherever you are."

"Muslim engineers have developed a bunch of cool devices for orthopraxy's sake."

Leor Halevi, a Vanderbilt University scholar who studies Islam, notes that there has been a marked increase in patents related to religious observance during the past decade. "Material objects have been part and parcel of Muslim rites from the origins of Islam. But in the past few decades," he argues, "Muslim engineers have developed a bunch of cool devices for orthopraxy's sake."

This Islamic tech boom included not only qibla finders but also a variety of gadgets that guide Muslims throughout the ritual of salat. One example, shown at the opening of this essay, is the "Islamic iPrayer," a digital prayer mat invented by Munif Zia. Zia's machine resembles a typical machine-woven prayer mat, but it has one additional feature: a digital audio console that is programmed to play the correct combination of recitations for each of the five daily prayers.

A child of immigrants, Zia developed the Islamic iPrayer to take the guesswork out of salat, a religious obligation that can be especially difficult for non-Arabic speakers. In a 2014 interview, Zia said:

There's a generation out there who want to pray but don't know how. They're embarrassed to go to the scholars of the mosque, who would say, "Look at your age, why don't you know how?" or to go to their parents, who often assume their children learned long ago.

The Islamic iPrayer has rows of buttons that correspond to each component of salat. As the user prays, they can see these buttons light up to indicate the next step for prayer. If this guidance isn't enough, they can also press each button to play audio clips of the correct Qur'anic recitation for each sequence. The high-tech prayer mat was created, in particular, for a generation of Muslims who might be struggling to learn traditional practices in an age full of competing stimuli. It also promises to educate without judgment and assures that one's daily prayers are being performed without error.

According to Halevi, devices like the Islamic iPrayer reflect an "anxiety about correct performance [that] is a characteristic of the religious life of modern Muslims." In an increasingly globalized and interconnected world, more and more Muslims come to prayer without the cultural and linguistic fluencies that can help make salat second nature.

Responding to a recent Pew survey, a majority of Muslim Americans (56 percent) said that most Muslims who come to the United States seek to adopt American customs and ways of life. Nearly two-thirds of Muslim Americans see no conflict between being a devout Muslim and living in a modern society. First-generation Muslim Americans come from a wide range of regions across the globe, including the Middle East, North Africa, South Asia, and Europe.

"Among the roughly one-in-five Muslim Americans whose parents also were born in the US," Pew reports, 59 percent are African Americans, most of whom have converted to Islam. These demographic changes help paint a picture of an intellectually diverse, multicultural group of practitioners who may look to technology for help with their unique spiritual needs.

Another device, which Halevi likens to a "Muslim Xbox," seems to take the technological mediation of prayer to its furthest extreme. The patented "interactive prayer machine" has three main components, says Halevi:

First, it has a prayer rug equipped with force sensors and vibrating motors. This pressure-sensitive pad registers when and in precisely what order a user's knees, hands and forehead touch the ground. Second, the system comes equipped with a camera that takes digital photographs of the user in motion. Through a posture detection technique involving the use of geometric modeling tools, a software program establishes a kinematic model of his or her bodily poses. Finally, this system also features a screen that coordinates the display of scriptural passages—in the original script of revelation or liturgy, as well as in transliteration and translation—with the performance of particular gestures of prayer.

The inventor of the device, Muslim engineer Wael Abouelsaadat, claims that many of today's Muslims are accustomed to the rapid pace of modern life and that they want to "customize their ritual experience with minimum time spent in educating themselves." By taking away the need for memorization, his invention seems to offer a form of machine-assisted presence. Like the Islamic iPrayer, the device aims to provide something that would otherwise require years of practice and memorization: a state of flow, where the prescriptions of prayer fade away into the cognitive background. For his part, though, Abouelsaadat does worry about "reducing prayer into a mere mechanical sequence of bodily gestures."

From the mechanical monk to the digital prayer mat, technology and prayer have long been intermixed. It is impossible to know what William James might say of the technologies of salat that came after him, but they seem to signify more than the "repetition of certain sacred formulae." Islamic prayer technology has helped new generations close the gap between traditional practices and modern life and, in some cases, seem to create meaningful variations of ritual and experience.

He is slain

Milton 25th October 1809.

The twelve children whose names are written on the back of this card were vaccinated by Doctr Amos Holbrook at the town inoculation in July last- they were tested by Small pox inoculation on the 10th Inst and discharged this day from the Hospital after offering to the world in the presence of most respectable witnesses who — honored Milton with their attendance on that occasion, an additional proof of the never failing power of that mild pre=ventive the Cow pock, against Small pox infection. a blessing great as it is singular in its kind, whereby the hearts of men ought to be elevated in praise to the — Allmighty Giver

Oliver Houghton

Chairman of the Committee for Vaccination

Conclusion

As this book neared completion in the first weeks of 2021, the United States approached the end of the first year of the COVID-19 pandemic. Since the diagnosis of a patient in the state of Washington with novel coronavirus on January 20, 2020, the entire country experienced fear of infection and hope for a cure, tensions surrounding and protests against lockdowns, and, most pervasively, enduring sadness at the loss of hundreds of thousands of lives. Some Americans dutifully wore face masks to curtail infection rates, while others resisted them. The diligent strove to find a new normal in the ongoing negotiation of social distancing and greater reliance on devices—smart phones, laptops, tablets, all deploying with an ever-expanding universe of apps—to bridge the gaps between themselves and others. All the while, the significance of intersections between religion, science, and technology was never more apparent.

While initial opposition to vaccination was largely religious in nature, in time the procedure was seen as a gift from "the almighty Giver," as seen on this 1809 notice from Milton, Massachusetts.

The COVID-19 era has provided the subject explored through *Discovery and Revelation* unwished-for relevance. It was a stark reminder that the questions addressed in this book, as well as in the exhibit that inspired it, are not matters of abstraction or theory but the very stuff of human lives. The inquiries with which we began—*Who are we? Where are we?* and *How should we live?*—took on new poignancy and urgency as hardships both monumental and mundane led many to reevaluate the answers they previously might have given to each.

Much as Cotton Mather sought to reconcile traditional religious views concerning sickness, healing, and repentance with the medical advancement that came to be known as inoculation, responses to the pandemic included religious resistance to the development of a vaccination program that promised to save countless lives. Faith, many behind the so-called anti-vax movement alleged, is the ultimate protection. One viral photograph taken at an anti-shutdown rally in

Harrisburg, Pennsylvania, captured this confusion of public health and personal devotion succinctly: A protester had painted the hood of his truck with the motto "Jesus is my vaccine." Fortunately, in both the eighteenth century and the twenty-first, more level heads prevailed—but not before it was shown that science and religion remain entwined in ways that affect us all.

This also has been a moment in which scientific opinion has proved as divisive for some religious communities as the early days of the evolution debate. "Listen to the scientists" became a mantra for those who wanted to limit in-person meetings to combat infection, while "church is essential" served as a rebuttal for those who insisted the freedom of religion ought to offer exemptions to houses of worship hoping to gather their flocks. Just as in the wake of Darwin, some religious

The back of the card from the preceding pages displays the names of twelve Massachusetts children vaccinated against smallpox.

Joshua Briggs
Samuel Alden
Thomas Street Briggs
Benjamin Church Briggs
Martin Briggs
George Briggs
Charles Briggs
John Smith
Catharine Bent
Sussanna Bent
Ruth Porter Horton
Mary Ann Belcher

communities adapted their traditions in the face of new circumstances and information while others became more strident in response. Yet for every church that refused to close temporarily, putting its members at risk, there were others that launched education campaigns and treated basic health protocols as holy writ.

Through it all, a variety of technologies transformed the terms of engagement between religion and science. Worship services moved online for millions. "Zoom church"—named for the popular service that put virtual meetings, once the exclusive province of the business world, in the hands of anyone with an internet connection—became a phenomenon in its own right, with rabbis self-broadcasting sermons, Zen teachers leading remote meditation, and gospel choirs joining each other to share hymns from the sanctuaries of their own homes. Just as radio had done for earlier generations of religious Americans, technology altered the terms of what it meant to gather together, perhaps in ways that will continue until the pandemic becomes a distant memory.

For religious Americans, technology has altered the terms of what it means to gather together.

One of the greatest insights offered by the study of history is that few circumstances are truly unprecedented, though that word has been used so often to describe the very recent past. When we look across time to consider earlier events that might help us make sense of the challenges we face, it is encouraging to realize how often they have been encountered and overcome. Bleak though it often is, history can also give us faith.

If there is a single faith that underlies the stories presented in this book, it is not faith in a deity, nor faith grounded in a particularly spiritual tradition, but faith simply in the human capacity to understand. In a nation composed of more than 300 million perspectives, Americans will never come to any grand agreement about the nature of the intersection of religion, science, and technology, but we may perhaps come to recognize that we have always sought answers. Sometimes they have arrived slowly, other times in a flash of insight as bright as one of Benjamin Franklin's lightning strikes, but always influenced by the complex interactions of the ideas all around us—scientific, religious, and myriad combinations of the two.

Notes

Introduction

"sameness of the electric fluid": Joseph Priestley, *The History and Present State of Electricity, with Original Experiments* (London: J. Dodsley, J. Johnson, B. Davenport, and T. Cadell, 1767), 179.

"When the Rain": Benjamin Franklin, "The Kite Experiment, 19 October 1752," *Founders Online*, National Archives, https://founders.archives.gov/documents/Franklin/01-04-02-0135.

"It is as impious": Jean-Antoine Nollet, quoted in Philip Dray, *Stealing God's Thunder: Benjamin Franklin's Lightning Rod and the Invention of America* (New York: Random House, 2005), 96.

"This Invention of Iron Points": John Adams, "Marginalia in Winthrop's Lecture on Earthquakes, December 1758," *Founders Online*, National Archives, https://founders.archives.gov/documents/Adams/01-01-02-0003-0003-0002.

"God's little workshop": George Washington Carver to Jack Boyd, quoted in Gary R. Kremer, *George Washington Carver: In His Own Words* (Columbia: University of Missouri Press, 1987), 135.

"The point is": Barbara McClintock, quoted in Nathaniel C. Comfort, *The Tangled Field: Barbara McClintock's Search for the Patterns of Genetic Control* (Cambridge, MA: Harvard University Press, 2009), 268.

"Does science sometimes conflict": "Religion and Science," Pew Research Center, Washington DC, October 22, 2015, https://www.pewresearch.org/science/2015/10/22/science-and-religion/.

Part 1: Revolutions

The Mechanical Monk

"self-acting": Elizabeth King, "Clockwork Prayer: A Sixteenth-Century Mechanical Monk," *Blackbird* 1, no. 1 (Spring 2002), https://blackbird.vcu.edu/v1n1/nonfiction/king_e/prayer_print.htm.

"the work of a master mechanic": Ibid.

"Between the sixteenth": Randall Styers, *Making Magic: Religion, Magic, and Science in the Modern World* (New York: Oxford University Press, 2004), 4.

"dubious encrustations": Ibid., 5.

And Yet It Moves

"uneven, rough, and crowded": Galileo Galilei, quoted in Andrea Frova and Mariapiera Marenzana, *Thus Spoke Galileo: The Great Scientist's Ideas and Their Relevance to the Present Day* (Oxford: Oxford University Press, 2006), 162.

"E pur si muove": Galileo Galilei, quoted in Giuseppe Baretti, *The Italian Library, Containing an Account of the Lives and Works of the Most Valuable Authors of Italy* (London: A. Millar, 1757), 42.

"The water clock": Galileo Galilei to Nicolas-Claude Fabri de Peiresc, 1635, ibid.

"more remote": Ibid.

"It is easy to see": Ibid.

Something of Ye Small-Pox

"Let us proceed": Cotton Mather, *The Christian Philosopher: A Collection of the Best Discoveries in Nature, with Religious Improvements* (Charlestown, MA: J. McKown, 1815), 21–22.

"great and visible decay": Cotton Mather, May 27, 1725, quoted in *Papers Relating to the History of the Church in Massachusetts, 1676–1785* (1873), 172.

"venomous, contagious": Cotton Mather, *Diary of Cotton Mather, 1681–1708*, ed., William Stevens Perry (Boston: Massachusetts Historical Society, 1911), 451.

"I had from a servant": Cotton Mather to John Woodward, July 12, 1716, quoted in Margot Minardi, "The Boston Inoculation Controversy of 1721–1722: An Incident in the History of Race." *William and Mary Quarterly* 61, no. 1 (2004): 47.

"undergone an Operation": Ibid.

"Negroish": William Douglass, *Inoculation of the Small Pox as Practised in Boston*, (Boston: J. Franklin, 1722), ibid., 66.

"It is not the healthy": John Williams, *Several Arguments, Proving, that Inoculating the Small Pox is Not Contained in the Law of Physick, either Natural or Divine, and Therefore Unlawful* (Boston: J. Franklin, 1721), 1.

"Mather, you dog!": Abijah Perkins Marvin, *The Life and Times of Cotton Mather* (Boston: Congregational and Sunday-School Publishing Society, 1892), 48.

"Is it not taking God's work out of his hands?": Samuel Grainger, *The Imposition of Inoculation as a Duty Religiously Considered* (Boston: N. Boone and J. Edwards, 1721), 12.

A Pharmacist's Herb Garden

"the first woman pharmacist": Nancy Brister, "Sister Xavier's Herb Garden at the Historic Old Ursuline Convent," *The Past Whispers* (website), http://www.thepastwhispers.com/NO_SisterXavier.html.

"The herbs that can be grown": Ibid.

"She had a zeal": Emily Clark, *Masterless Mistress: The New Orleans Ursulines and the Development of a New World Society, 1727–1834* (Chapel Hill: University of North Carolina Press, 2012), 96–100.

"She gave herself": Ibid.

"sacred and inviolate": Thomas Jefferson, "From Thomas Jefferson to Ursuline Nuns of New Orleans, 13 July 1804," *Founders Online*, National Archives, https://founders.archives.gov/documents/Jefferson/99-01-02-0068.

Electric Fire

"lightning pictures": Dray, *Stealing God's Thunder*, 65.

"diabolical agency": Ibid., 66.

"impious": Ibid., 96.

"It has pleased God": Benjamin Franklin, "Poor Richard Improved, 1753," *Founders Online*, National Archives, https://founders.archives.gov/documents/Franklin/01-04-02-0148.

"speaks as if he thought it Presumption": Benjamin Franklin, "From Benjamin Franklin to Cadwallader Colden, 12 April 1753," *Founders Online*, National Archives, https://founders.archives.gov/documents/Franklin/01-04-02-0167.

"it is not Lightning": Ibid.

"that it is possible": John Winthrop, *A Lecture on Earthquakes* (Boston: Edes & Gill, 1755), 37.

Thomas Paine's Clockwork Universe

"a machinery of clock-work": Thomas Paine, *The Age of Reason: Being an Investigation of True and Fabulous Theology* (New York: D. M. Bennett, 1877), 40.

"the infinity of space": Ibid.

"the eternal evidence": Ibid.

"The Bible of Creation": Thomas Paine, *The Age of Reason, Part the Second: Being an Investigation of True and Fabulous Theology* (London: Cock and Swine, 1796), 83.

"a revolution in the system of religion": Ibid.

An Almanac of Strange Dreams

"the blacks . . . are inferior": Thomas Jefferson, *Notes on the State of Virginia* (Boston: Lilly and Wait, 1832), 150.

"that one universal Father": Benjamin Banneker, "To Thomas Jefferson from Benjamin Banneker, 19 August 1791," *Founders Online*, National Archives, https://founders.archives.gov/documents/Jefferson/01-22-02-0049.

"Having long had unbounded desires": Benjamin Banneker in "To Thomas Jefferson from Benjamin Banneker, 19 August 1791," *Founders Online*, National Archives, https://founders.archives.gov/documents/Jefferson/01-22-02-0049.

"a document to which your whole colour": Thomas Jefferson in "From Thomas Jefferson to Benjamin Banneker, 30 August 1791," *Founders Online*, National Archives, https://founders.archives.gov/documents/Jefferson/01-22-02-0091.

"December 13, 1797": Benjamin Banneker, quoted in Silvio Bedini, *The Life of Benjamin Banneker* (New York: Scribner, 1972), 334.

"April 24, 1802": Ibid., 335.

"What are these dreams": Matthew Gilmore, "The Dreams of Benjamin Banneker," *Maryland Historical Society*, https://www.mdhistory.org/the-dreams-of-benjamin-banneker/

Religious Freedom and the Air We Breathe

"Antagonists think they": Thomas Jefferson, quoted in John Bernard, "Recollections of President Jefferson," in *Jefferson in His Own Time*, ed. Kevin J. Hayes (Iowa City: University of Iowa Press, 2012), 112.

"breast felt peculiarly": Joseph Priestley, quoted in Steven Johnson, *The Invention of Air* (New York: Riverhead, 2008), 99.

"happiest": Ibid., 55.

"generally called the heterodox side": Joseph Priestley, quoted in Dan Eshet, "Rereading Priestley: Science at the Intersection of Theology and Politics," *History of Science* 39, no. 2 (2001): 134.

"labyrinths of error": Joseph Priestley, *An Essay on the First Principles of Government* (London: J. Johnson, 1768), 261.

"I find it absolutely impossible": Joseph Priestley, *Experiments and Observations on Different Kinds of Air and Other Branches of Natural Philosophy* (Birmingham: Thomas Pearson, 1790), 1:xviii.

Part 2: Evolutions

The Bible Surgeon

"excellent pedestrian": John Cory, *The Life of Joseph Priestley* (Birmingham: Wilks, Grafton & Co., 1804), 44.

"over and over again": Thomas Jefferson, "Thomas Jefferson to John Adams, 22 August 1813," *Founders Online*, National Archives, https://founders.archives.gov/documents/Jefferson/03-06-02-0351.

"No problem was too abstruse": William Eleroy Curtis, *The True Thomas Jefferson* (Philadelphia: Lippincott, 1901), 346.

"Jesus did not mean": Thomas Jefferson, "From Thomas Jefferson to William Short, 4 August 1820," *Founders Online*, National Archives, https://founders.archives.gov/documents/Jefferson/98-01-02-1438.

"the result of a life of inquiry": Thomas Jefferson, "From Thomas Jefferson to Benjamin Rush, 21 April 1803," *Founders Online*, National Archives, https://founders.archives.gov/documents/Jefferson/01-40-02-0178-0001.

"Uniformity of opinion": Thomas Jefferson, *Notes on the State of Virginia* (Boston: Lilly and Wait, 1832), 167.

"of a sect by myself": Thomas Jefferson, "Thomas Jefferson to Ezra Stiles Ely, 25 June 1819," *Founders Online*, National Archives, https://founders.archives.gov/documents/Jefferson/03-14-02-0428.

"that there is no God": Thomas Jefferson, "From Thomas Jefferson to Peter Carr, with Enclosure, 10 August 1787," *Founders Online*, National Archives, https://founders.archives.gov/documents/Jefferson/01-12-02-0021.

A Family Affair

"Geology is usually regarded": Edward Hitchcock, *The Religion of Geology and its Collected Sciences* (Boston: Phillips, Sampson, and Co., 1856), 1.

"Men of respectable ability": Ibid., 2.

"prejudices . . . striking misapprehensions": Ibid.

"While I have described": Ibid., iii.

What Hath God Wrought

"illustrating the providence": Joseph A. Copp, *A Discourse Preached in the Broadway Church, Chelsea* (Boston: T. R. Marvin & Son, 1858).

"The improvements of art": Ibid.

"heathen minds": Jenna Supp-Montgomerie, *When the Medium Was the Mission: The Atlantic Telegraph and the Religious Origins of Network Culture* (New York: New York University Press, 2021), 48.

"I received letters": Leah Fox, "Certificate of Mrs. Margaret Fox, Wife of John D. Fox, the Occupant of the House," in Ann Leah Underhill, *The Missing Link in Modern Spiritualism* (New York: Thomas R. Knox & Co., 1885), 49.

"the relation of the facts": Henry Calderwood, *The Relations of Science and Religion: The Morse Lecture, 1880* (New York: Robert Carter & Bros., 1881), v.

Our Celestial Visitant

"Mrs. M. Baker": "Ladies Department." *Daily Republican* (Monongahela, PA), December 5, 1883.

"fine gauze wavering in a strong breeze": Étienne Léopold Trouvelot, "Undulations Observed in the Tail of Coggia's Comet, 1874," *Proceedings of the American Academy of Arts and Sciences*, vol. 13 (May 1877–May 1878): 185.

"Like the shepherds of yore": Ben Hur Wilson, "The Great Comet of 1882," *The Palimpsest* 21, no. 9 (September 1940): 286.

"The antics of the comet": Ibid., 293.

"a separation was seen": Ibid.

"present comet in the Eastern sky": Ibid., 293–94.

"To Our Celestial Visitant": Ibid., 295.

The Anatomy of the Soul

"bundles of threads": Franz Joseph Gall, quoted in John van Wyhe, "Franz Joseph Gall," *The History of Phrenology on the Web* (website), accessed February 11, 2021, http://www.historyofphrenology.org.uk/fjgall.html.

"sense of God and Religion": Franz Joseph Gall, quoted in John van Wyhe, "The Phrenological Organs," *The History of Phrenology on the Web* (website), accessed February 11, 2021, http://www.historyofphrenology.org.uk/organs.html.

"the first principles of morality": Francis II, quoted in John Van Wyhe, "The Authority of Human Nature: The 'Schädellehre' of Franz Joseph Gall," *The British Journal for the History of Science* 35, no. 1 (2002): 25.

"truth of phrenology": Minna Scherlinder Morse, "Facing a Bumpy History," *Smithsonian Magazine*, October 1997, https://www.smithsonianmag.com/history/facing-a-bumpy-history-144497373.

"very large": James A. Garfield, quoted in Mary Lintern, "Phrenology in Victorian America," *The Garfield Observer* (blog), August 2012, accessed December 15, 2020, https://garfieldnps.wordpress.com/2012/08/31/phrenology-in-victorian-america/.

"Destructiveness": Morse, "Facing a Bumpy History."

"If revelation and phrenology": Orson Squire Fowler and Lorenzo Niles Fowler, *Phrenology: Proved, Illustrated, and Applied* (New York: W. H. Colyer, 1836), 124.

"Study and admire": Orson Squire Fowler and Lorenzo Niles Fowler, *New Illustrated Self-Instructor in Phrenology and Physiology* (London: W. Tweedie, 1867), 126.

Darwin in America

"The green and budding twigs": Charles Darwin, *On the Origin of Species by Means of Natural Selection* (London: Murray, 1859), 129–30.

"thought of God": Louis Agassiz, quoted in David Starr Jordan, "Louis Agassiz, Teacher," *The Scientific Monthly* 17, no. 5 (1923): 408.

"Darwinian teleology": Asa Gray, *Darwiniana: Essays and Reviews pertaining to Darwinism* (New York: Appleton, 1876), 322.

"divine desire to know": Ibid., 94.

"We may take it to be": Asa Gray, *Natural Science and Religion: Two Lectures Delivered to the Theological School of Yale College* (New York: Scribner's, 1880), 8.

"I cannot honestly go as far as you": Charles Darwin, "To Asa Gray, 5 September [1857]," *Darwin Correspondence Project*, https://www.darwinproject.ac.uk/letter/DCP-LETT-2136.xml.

"like the annihilation": Asa Gray, quoted in Janet Browne, "Asa Gray and Charles Darwin: Corresponding Naturalists," *Harvard Papers in Botany* 15, no. 2 (2010): 209.

"We cling to a long-accepted theory": Asa Gray, "Darwin on The Origin of Species: A Book Review," *The Atlantic* 6, no. 33 (July 1860): 109–16, https://www.theatlantic.com/ideastour/science/gray-full.html.

Quilting the Cosmos

"Adam and Eve": Kyra E. Hicks, *This I Accomplish: Harriet Powers' Bible Quilt and Other Pieces* (Richmond, VA: Black Threads Press, 2009), 14. The complete description is as follows:

Top Row—left to right

1. *Adam and Eve naming the animals, including a camel, elephant, leviathan, and ostrich, in the Garden of Eden. Adam and Eve are also being tempted by a crawling serpent.*
2. *Adam and Eve with their son, Cain. A bird of paradise, made of red and green calico, is stitched in the bottom right corner.*
3. *The Devil with pink eyes stands surrounded by seven stars.*

Middle Row—left to right

4. *Cain killing his brother Abel, who is lying in a stream of his own blood. The sheep Cain is tending witness the attack.*

5. *Cain and his wife in the City of Nod. An orange calico lion stands among bears, an elk, leopards, and a "kangaroo hog."*

6. *Jacob sleeps while a winged angel seems to ascend or descend Jacob's ladder.*

7. *A dove from heaven flying over Jesus and John the Baptist after John baptized the Son of God.*

Bottom Row—left to right

8. *A bleeding Jesus crucified with two thieves. The circular objects above the crosses "represent the darkness over the earth and the moon turning into blood."*

9. *Judas Iscariot surrounded by thirty pieces of silver, his payment for betraying Jesus. The pieces of silver were originally made of green calico fabric. The largest circular object is a star "that appeared in 1886 for the first time in three hundred years."*

10. *Christ with his disciples at the Last Supper. According to [a] letter sharing Powers's own description, it is Judas Iscariot who is "clothed in drab, being a little off-color in character."*

11. *The Holy Family, Mary, Joseph, and Jesus, with the Star of Bethlehem shining over them. The crosses represent the burden Christ carried in life.*

"commenced to learn": Harriet Powers, quoted in Hicks, *This I Accomplish*, 37.

"She arrived one afternoon": Jennie Smith, quoted in Laurel Thatcher Ulrich, "'A Quilt unlike Any Other': Rediscovering the Work of Harriet Powers," in *Writing Women's History*, ed. Elizabeth Anne Payne (Jackson: University Press of Mississippi, 2011), 86–87.

"Harriet was one among": Ulrich, "A Quilt unlike Any Other," 93.

Conflict or Agreement?

"few men of science": American Chemical Society, "John W. Draper and the Founding of the American Chemical Society." American Chemical Society website, accessed December 17, 2020, https://www.acs.org/content/acs/en/education/whatischemistry/landmarks/draperacs.html.

"The history of Science is not a mere record": John William Draper, *History of the Conflict between Religion and Science* (New York: Appleton, 1878), vi.

"The antagonism": Ibid.

"unchangeable, stationary": Ibid., vii.

"Science . . . has never": Ibid., xi.

"the American gospel": Paul Blanshard, *American Freedom and Catholic Power* (Boston: Beacon Press, 1949), 211.

"In all modern history": Andrew Dickson White, quoted in Ronald L. Numbers, "Science and Religion," *Osiris* 1 (1985): 59.

"As a historical tool": Colin A. Russell, "The Conflict of Science and Religion," in *Science and Religion: A Historical Introduction*, ed. Gary B. Ferngren (Baltimore: John Hopkins University Press, 2002), 8.

"reinstate primitive Christianity": Mary Baker Eddy, *Manual of the Mother Church: The First Church of Christ Scientist in Boston, Massachusetts* (Boston: Allison V. Stewart, 1913), 17.

"Health is not a condition of matter": Mary Baker G. Eddy, *Science and Health with Key to the Scriptures* (Boston: Nixon, 1891), 14.

"If Christianity is not Scientific": Ibid., 288.

Part 3: Complexity

Her Heavenly Radium

"There must be": Marie Curie, *Pierre Curie, with Autobiographical Notes by Marie Curie*, trans. Charlotte and Vernon Kellogg (New York: Macmillan, 1923), 182.

"God is the perfect radium": Jay A. Ford, quoted in "Minister Refers to God As the Heavenly Radium" *St. Louis Republic*, January 11, 1904.

"We are constantly being reminded": Wallace Radcliffe, quoted in "'Spiritual Radium' is Subject of Sermon," *Washington Herald*, January 12, 1914.

"Radium is another revelation": May Barnard Wiltse, *Theorem, or Teleology of Spiritualism* (New Rockford, ND: Eddy County Provost, 1909), 57.

"I must admit": Pierre Curie, quoted in Anna Hurwic, *Pierre Curie*, trans. Lilananda Dasa and Joseph Cudnik (Paris: Flammarion, 1995): 66.

"As to the age": Frederick Boyd Stevenson, "Why All Conflict Should Cease Between Science and Religion," *Brooklyn Daily Eagle*, December 24, 1922.

Religion on the Radio

"Last night for the first time": Spencer Miller, "Radio and Religion," in "Radio: The Fifth Estate," special issue, *Annals of the American Academy of Political and Social Science* 177 (January 1935): 135–36.

"the largest primetime distribution": Steve Craig, *Out of the Dark* (Tuscaloosa: University of Alabama Press, 2009), 107.

"We do enjoy the broadcasts": Ibid., 108.

"I cook on a ranch": Ibid.

Trial of the Century

"It shall be unlawful": House Bill No. 185, 64th General Assembly (TN 1925), http://law2.umkc.edu/faculty/projects/ftrials/scopes/tennstat.htm.

"fine mess": John Scopes, quoted in Douglas O. Linder, "John Scopes," 2004, *Famous Trials* (website), accessed December 15, 2020, http://law2.umkc.edu/faculty/projects/ftrials/scopes/Sco_sco.htm.

"beginners in secondary school": George William Hunter, *A Civic Biology: Presented in Problems* (New York: American Book Company, 1914), 7.

"Evolution of Man": Ibid., 195.

"The Scopes trial, from the start": H. L. Mencken, "The Monkey Trial: A Reporter's Account," *Baltimore Evening Sun*, July 18, 1925.

"Oh, the folks in Tennessee": "The John Scopes Trial," lyrics by Carson Robison, quoted in Studs Terkel and Sydney Lewis, *Touch and Go* (New York: New Press, 2007), 50.

"The Races of Man": Hunter, *A Civic Biology*, 196.

"If such people": Ibid., 263.

"I don't know if I'm a Christian": Linder, "John Scopes."

"All of biology": Ibid.

Playing Dice with the Universe

"God does not play": Albert Einstein to Max Born, December 12, 1926, quoted in Max Born, *The Born-Einstein Letters, 1916–1955: Friendship, Politics, and Physics in Uncertain Times*, trans. Irene Born (New York: Walker and Company, 1971), 91.

"whether I would": Albert Einstein, quoted in John Brooke and Geoffrey Cantor, *Reconstructing Nature: The Engagement of Science and Religion* (New York: Oxford University Press, 1998), 227.

"The word God": Albert Einstein to Eric Gutkind, January 3, 1954, quoted in James Barron, "Einstein's 'God Letter,' a Viral Missive from 1954," *New York Times*, December 2, 2018.

"cosmic religious feeling": Albert Einstein, "Religion and Science," *New York Times Magazine*, November 9, 1930.

"irreconcilable antagonists": Albert Einstein, "Science and Religion," in *Ideas and Opinions* (New York: Crown, 1954), 40.

"science without religion is lame, religion without science is blind": Ibid.

Immortal Life?

"Her cells were part": Rebecca Skloot, *The Immortal Life of Henrietta Lacks* (New York: Random House, 2010), 4.

"Henrietta was chosen": Ibid., 295.

"Nelson wanted to create": Dorothy Moss, quoted in Ryan P. Smith, "Henrietta Lacks is Immortalized in Portraiture," *Smithsonian Magazine*, May 15, 2018, https://www.smithsonianmag.com/smithsonian-institution/famed-immortal-cells-henrietta-lacks-immortalized-portraiture-180969085/.

Earthrise

"something appropriate": Frank Borman, quoted in "Apollo 8: Christmas at the Moon," NASA website, December 23, 2019, https://www.nasa.gov/topics/history/features/apollo_8.html.

"We were told": Ibid.

"The first ten verses": Ibid.

"Around the world": Teasel Muir-Harmony, "How Apollo 8 Delivered Christmas Eve Peace and Understanding to the World," *Smithsonian Magazine*, December 11, 2020, https://www.smithsonianmag.com/smithsonian-institution/how-apollo-8-delivered-moment-christmas-eve-peace-and-understanding-world-180976431.

"seeking an order enjoining": O'Hair v. Paine, 312 F. Supp. 434 (W. D. Tex. 1969), https://law.justia.com/cases/federal/district-courts/FSupp/312/434/1468840/.

"The various plaintiffs": Ibid.

"With regard to the scheduling": Ibid.

Your Brain on God

"The human brain": George H. W. Bush, Proclamation No. 6158, (July 18, 1990).

"This had never been seen": Richard Davidson, quoted in Lauren Effron, "Neuroscientist Richie Davidson Says Dalai Lama Gave Him 'a Total Wake-Up Call' that Changed His Research Forever," *ABC News*, July 27, 2016.

"the use, possession": American Indian Religious Freedom Act Amendments of 1994, Pub. L. No. 103-344, 108 Stat. 3125 (1994), https://www.law.cornell.edu/uscode/text/42/1996a.

"remarkable mental and physical effects": Weston La Barre, David P. McAllester, J. S. Slotkin, Omer C. Stewart, and Sol Tax, "Statement on Peyote," Science 114, no 2,970 (1951): 582.

"opaque goggles": Leslie A. Ruttan, Michael A. Persinger, and Stanley Koren, "Enhancement of Temporal Lobe-Related Experiences during Brief Exposures to Milligauss Intensity Extremely Low Frequency Magnetic Fields," *Journal of Bioelectricity*, 9, no. 1 (1990): 33.

The Technology of Prayer

"the very soul and essence of religion": William James, *The Varieties of Religious Experience* (New York: Longmans, Green, 1902), 464.

"no vain exercise": Auguste Sabatier, quoted in ibid., 464.

"You're doing the same thing": Richard Warp, quoted in Amy Standen, "Brain Games: Move Objects with Your Mind to Find Inner Calm?" *NPR Morning Edition*, January 21, 2014, https://www.npr.org/sections/alltechconsidered/2014/01/21/263078049/brain-games-move-objects-with-your-mind-to-find-inner-calm.

"Material objects have been": Leor Halevi, quoted in Peter Manseau, "Will an Islamic Tech Boom Help Teach Young Muslims To Pray?" *Aljazeera America*, February 28, 2014, http://america.aljazeera.com/articles/2014/2/28/will-an-incipientislamictechboomteachyoungmuslimstopray.html.

"There's a generation out there": Munif Zia, quoted in ibid.

"anxiety about correct performance": Leor Halevi, "The Muslim Xbox," *Reverberations: New Directions in the Study of Prayer* (Social Science Research Council), May 31, 2013, http://forums.ssrc.org/ndsp/2013/05/31/the-muslim-xbox/.

"Among the roughly one-in-five": "Muslims Americans: No Signs of Grown in Alienation or Support for Extremism," Pew Research Center, Washington DC, August 30, 2011, https://www.pewresearch.org/politics/2011/08/30/muslim-americans-no-signs-of-growth-in-alienation-or-support-for-extremism/.

"interactive prayer machine": Halevi, "The Muslim Xbox."

"customize their ritual experience": Ibid.

"reducing prayer": Ibid.

Index

Page numbers in **bold** refer to images

Acknowledgments

Discovery and Revelation: Religion, Science, and Making Sense of Things would not have been possible without the support of H. Bruce McEver and the Foundation for Religious Literacy, whose visionary leadership has benefited so many engaged in the important work of improving the public understanding of religion.

Funding for the *Discovery and Revelation* exhibition at the National Museum of American History, upon which this book is based, was provided by Lilly Endowment Inc., with additional support from the John Templeton Foundation for a series of consultations with scholarly advisers, including Luis Campos, Francisca Cho, Edward Davis, Christian Goodwillie, John Haught, Edward Larson, John Lardas Modern, Ahmed Ragab, Mary-Jane Rubenstein, Lily Santoro, Lea Schweitz, Myrna Perez Sheldon, Randall Styers, Jenna Supp-Montgomerie, Hava Tirosh-Samuelson, Frederick Ware, and Judith Weisenfeld. Our gratitude to all for helping this project take shape.

Though only two names appear on the cover of this book, it owes its existence equally to Lauren Safranek and her tireless efforts developing the exhibition while researching and acquiring the images that appear throughout these pages. Heartfelt thanks also go to our many colleagues who made vital contributions to the creation of the *Discovery and Revelation* book and exhibition, including Anthea Hartig, Benjamin Filene, Howard Morrison, Heidi Helgerson, Amanda Bowen, Scott Nolley, Lily Hoffman, Margaret Grandine, Katharine Klein, Cassie Mancer, Jaclyn Nash, Bennie Brunton, Valeska Hilbig, Fiona Meagher, and Ushonda Holmes.

The team brought together by Smithsonian Books—Carolyn Gleason, Jaime Schwender, Julie Huggins, Matt Litts, Sarah Fannon, and Tom Fredrickson, along with designers Antonio Alcalá and Marti Davila from Studio A—have all enhanced and improved this work significantly.

Just as the interactions between religion and science throughout history have never occurred in isolation but always within communities, this book has been informed by many voices. Any mistakes that may be found are, of course, our own.

Peter Manseau and Andrew Ali Aghapour

Illustration Credits

2: NASA; **6**: Philadelphia Museum of Art: Gift of Mr. and Mrs. Wharton Sinkler, 1958, 1958-132-1; **9**: Photo by Richard W. Strauss, National Museum of American History, Smithsonian Institution; **10**: Photo by Jaclyn Nash, National Museum of American History, Smithsonian Institution. Courtesy of the Carnegie Institute of Washington; **16, 18**: Photo by Jaclyn Nash, National Museum of American History, Smithsonian Institution; **21**: National Museum of American History & Smithsonian Museum Conservation Institute; **22, 24**: Courtesy of Smithsonian Libraries and Archives, Washington, DC; **27**: Cooper Hewitt, Smithsonian Design Museum. Gift of Eleanor and Sarah Hewitt; **28**: National Library of Medicine, National Institutes of Health; **31**: Library of Congress, Prints and Photographs Division, LC-USZC4-4597; **32, 35**: Courtesy of Ursuline Convent New Orleans Archive & Museum; **36**: Historical and Interpretive Collections of The Franklin Institute, Philadelphia, PA; **39**: Library of Congress, Prints and Photographs Division, LC-USZ62-90398; **40**: Derby Museum and Art Gallery, Derby Museums; **42–43, 44**: National Museum of American History & Smithsonian Institution Archives; **45**: Library of Congress, Prints and Photographs Division, LC-DIG-ppmsca-24327; **46**: Courtesy of the Maryland Center for History and Culture; **49**: Black Heritage: Benjamin Banneker and Banneker as Surveyor © 1980 United States Postal Service®. All Rights Reserved. Used with Permission.; **50, 52**: National Museum of American History, Gift of Miss Frances D. Priestley; **53**: National Museum of American History & Smithsonian Institution Archives. Courtesy of the American Chemical Society; **54**: National Museum of American History, Smithsonian Institution; **56, 74**: National Museum of American History & Smithsonian Institution Archives. Gift of Patricia Hill McCloy and Kathryn Hill Meardon; **58, 60, 61**: Photo by Hugh Talman, National Museum of American History, Smithsonian Institution; **62**: Library of Congress, Prints and Photographs Division, LC-USZC4-2474; **64, 66–67, 68, 69*t*, 69*b***: Amherst College Archives & Special Collections; **70**: Photo by Jaclyn Nash, National Museum of American History, Smithsonian Institution. Gift of Edward Lind Morse; **72**: National Museum of American History, Smithsonian Institution. From Western Union Telegraph Co.; **73**: Library of Congress, Prints and Photographs Division, LC-DIG-ppmsca-51817; **76**: Smithsonian National Air and Space Museum (NASM 9A14673); **78**: *The Household Physician*, 1905. Flickr, image courtesy of William Cresswell. https://www.flickr.com/photos/crackdog/14866775264; **81**: National Museum of American History, Smithsonian Institution; **84**: National Museum of American History, Smithsonian Institution; **87**: Courtesy of the Smithsonian Libraries and Archives, Washington, DC; **88*l***: Bettmann / Contributor via Getty Images; **88*m***: *Hebrew Union College, and Other Addresses* (Cincinnati, OH: Ark Publishing Co., 1916) by Kaufmann Kohler: https://archive.org/details/hebrewunioncolleookohliala; **88*r***: Courtesy of the Library of Congress, Prints and Photographs Division, LC-USZ62-28482; **90, 92, 93**: National Museum of American History & Smithsonian Institution Archives. Gift of Mr. and Mrs. H. M. Heckman; **94**: National Museum of American History, Smithsonian Institution; **97**: © The Metropolitan Museum of Art. Image Source: Art Resource, NY; **98**: Library of Congress, Prints and Photographs Division, LC-USZ62-59770; **100, 112–113**: Photograph by Watson Davis, black and white photographic print, Smithsonian Institution Archives, Image ID# SIA2007-0124; **102**: Library of Congress, Prints and Photographs Division, LC-USZ62-91224; **105**: Courtesy of the National Museum of Nuclear Science & History; **106**: National Museum of American History & Smithsonian Institution Archives. Gift of Gertrude Maud Goldsmith; **108**: National Museum of American History, Smithsonian Institution. Gift of the Church of the Covenant; **110**: Photograph by Watson Davis, black and white photographic print, Smithsonian Institution Archives, Image ID# SIA2008-1121; **114–115**: Photograph by Watson Davis, black and white photographic print, Smithsonian Institution Archives, Image ID# SIA2007-0112; **118**: Library of Congress, Prints and Photographs Division, LC-USZ62-106042; **121**: National Museum of American History & Smithsonian Institution Archives; **122, 124**: Collection of the Smithsonian National Portrait Gallery and National Museum of African American History & Culture, Gift from Kadir Nelson and the JKBN Group, LLC. © 2017 Kadir Nelson; **126**: Courtesy of Adler Planetarium, Chicago, Illinois; **128, 129**: NASA; **130**: © Cary Wolinsky 2003; **132**: Photo by NMAI Photo Services. National Museum of the American Indian, Smithsonian Institution, 25/3662; **133**: Courtesy of Rick Chard; **134**: Photo by Jaclyn Nash, National Museum of American History, Smithsonian Institution. Property of Peter Manseau; **135**: Courtesy of EMOTIV and Andrew Miller, Awesome Photography; **136**: National Museum of American History & Smithsonian Institution Archives; **140, 142**: National Museum of American History, Smithsonian Institution; **160**: Courtesy of *The Press of Atlantic City*, photograph by Edward Lea.

Succeeding page: The intersections of religion and science during the COVID-19 pandemic could be seen in the addition of protective masks to religious symbols, such as this statue of Jesus at Our Lady Star of the Sea Church in Atlantic City, New Jersey.